Environmental Soil Science

A.K. Singh
Program Coordinator
Krishi Vigyan Kendra (ICAR)
Birchandramanu, P.O. Manpathar,
Santirbazar, South Tripura

IBDC

International Book Distributing Co.
(Publishing Division)

Published by

INTERNATIONAL BOOK DISTRIBUTING CO.
(Publishing Division)
Khushnuma Complex Basement
7, Meerabai Marg (Behind Jawahar Bhawan),
Lucknow 226 001 U.P. (INDIA)
Tel. : 91-522-2209542, 2209543, 2209544, 2209545
Fax : 0522-4045308
E-Mail : ibdco@airtelmail.in

First Edition 2009

Rs-1100

ISBN 81-8189-279-8

Composed & Designed at :

Panacea Computers
3rd Floor, Agarwal Sabha Bhawan, Subhash Mohal
Sadar Cantt. Lucknow-226 002
Tel : 0522-2483312, 9335927082, 9452295008
E-mail : prasgupt@rediffmail.com

Printed at:

Salasar Imaging Systems
C-7/5, Lawrence Road Industrial Area
Delhi - 110 035
Tel. : 011-27185653, 9810064311

Dr. Biplab Bhusan Basu
Director
School of Fundamental Research

Foreword

Soil along with water, is a finite primary resource for the sustenance of life and particularly that of the human society in general on earth and as such it is more important than ever before because the soils of world with unprecedented erosion due to developmental we have chosen for faster growth, are non-renewable over the human time-frame and liable to misuse and mismanagement. Soil is no doubt a natural sink and a biochemical reactor but its capacity and ability is limited to absorb, fix and detoxify most of the pollutants.

This book discusses the research-based knowledge about interaction between soils and environmental disturbances in a lucid and coherent manner. The important feature of the book is that it provides information on various soil conditions in relation to plant growth, its chemistry, cascading effect of farm inputs on soil and water quality and implication of these effects on agricultural and environmental sustainability.

I hope this book should provide good foundation in environmental soil science to students, researchers and academician of applied soil science and environmental sciences, agriculture science, plant science, and others and inspire them to take up further studies in the subject.

Kolkata, India **Dr. Biplab Bhusan Basu**

Preface

Environmental science has grown rapidly in the last few decades and is now a major research and educational focus at universities around the world. Because of the multidisciplinary nature of environmental science, one cannot be expected to be an expert in all topics encompassing this extensive field. It is therefore important to have a book that deals with specific environmental science topics. *Environmental Soil Science* presents the fundamental concepts of soil science and applies them to environmentally significant reactions in soil.

This book is divided into 14 chapters. Chapters cover basic information on soil conditions and environmental plant growth, plant nutrients, soil-plant-water relationship, and soil quality useful to students new to the discipline, including importance of acid- and salt-affected environments, and environmental concern with fertilizer use. However, discussion also extends to heavy metals, arsenic contamination, acid mine drainage, toxic gaseous emissions, agrochemicals, radionuclides, and untreated wastewater irrigation as well as approaches for remediation of contaminated soils. Numerous figures, tables, and equations clarify the material discussed throughout the book. This book will be of interest to not only undergraduate and postgraduate students of applied soil science and environmental sciences, but also in natural resources, plant science, agronomy, engineering, geology, botany, geography, landscape ecology, and others.

The author duly acknowledged the sources of the diagrams and tables that have been reproduced from books and other publications.

A. K. Singh

Acronyms and abbreviations used in text

ADP	adenosine diphosphate
AMD	acid mine drainage
As	arsenic
ATP	adenosine triphosphate
AWC	available water capacity
AWHC	available water holding capacity
BHC	benzene hexachloride
Bq	becquerel (unit of radioactivity in place of 'curie')
DAP	diammonium phosphate
DDT	di-chloro diphenyl trichloroethane
2,4-D	2,4- dichlorophenoxy acetic acid (herbicide)
DTPA	diethylne triamine penta acetic acid
EC	electrical conductivity
ECCE	effective calcium carbonate equivalent
EDDHA	ethylene diamine dihydroxyphenyl acetic acid
EDTA	ethylene diamine tetra acetic acid
EGTA	ethylene glycol tetra acetic acid
EPA	Environmental Protection Agency
ESP	exchangeable sodium percentage
ET	evapotranspiration
ETc	crop evapotranspiration
FAO	Food and Agriculture Organisation of the United Nation
FC	field capacity
HEDTA	hydroxyethyl ethylene diamine tri acetic acid
IAA	indol acetic acid
KCl	potassium chloride
MAP	monoammonium phosphate
MBC	microbial biomass carbon
MMA	monomethyl arsonic acid
MOP	muriate of Potash
MRG	maximum root growth
NTA	mitrilo tri acetic acid
Pa	pascal unit
PGPR	plant growth promoting rhizobacteria
PGR	plant growth regular
PMN	potentially mineralzable nitrogen
POM	particulate organic matter
PWP	permanent wilting point
SAR	sodium absorption ratio
SOM	soil organic matter
SSP	single super phosphate
TOC	total organic carbon
TSP	triple superphosphate
2,4,5-T	2,4, 5- triichlorophenoxy acetic acid (herbicide)
USDA	United States Department of Agriculture
USEPA	United States Environment Protection Agency
VAM	vesicular arbiscular mycorrhiza
VOSC	volatile organic sulphate compound
WHC	water holding capacity
WHO	World Health Organisation of the United Nation

Contents

1

Soil and Soil Productivity

Soil is perhaps the most important natural resource. It is important to all, for without soil there would be no life on Earth. Our food and much of our clothing and shelter come from the soil. Soil supports the gigantic agricultural system which is the major contributor to the country's development and continued prosperity.

Some soils are extremely productive while others are not so productive. Each soil has a set of unique characteristics which distinguishes it from other soils. These characteristics determine the potential productivity of the soil. Soil productivity is a result of how well the soil is able to receive and store moisture and nutrients as well as providing a desirable environment for all plant root functions.

WHAT IS SOIL?

Soil is the unconsolidated mineral and organic material on the immediate surface of the Earth which provides nutrients, moisture, and anchorage for land plants. The principal components of soil are mineral material, organic matter, water and air. These are combined in widely varying amounts in different soils. In a typical loam soil, solid material and pore space are equally divided on a

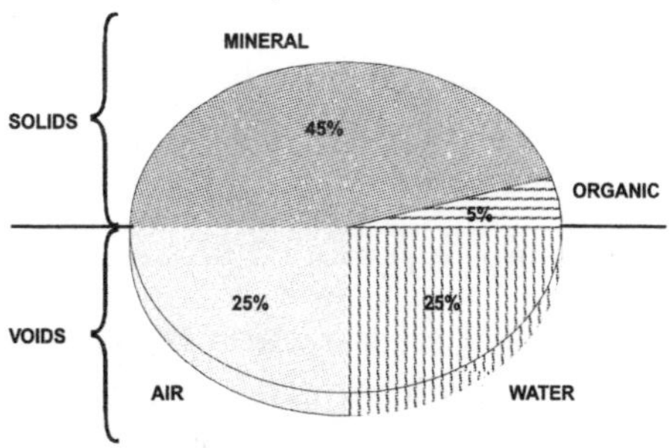

Figure 1.1. Volume composition of a desirable surface soil.

volume basis, with the pore space containing nearly equal amounts of water and air. The approximate proportions are illustrated in Figure 1.1.

HOW SOILS ARE FORMED?

The development of soils from parent rock is a long term process involving physical and chemical weathering along with biological activity. The wide variety of soils and their properties are a function of the soil forming factors including parent material, climate, living organisms, topography and time.

The initial action on the parent rock is largely mechanical i.e. cracking and chipping due to temperature changes. As the rock is broken, the total surface area exposed to the atmosphere increases. Chemical action of water, oxygen, carbon dioxide and various acids further reduce the size of rock fragments and change the chemical composition of many resulting particles. Finally, the microorganism activity and higher plant and animal life contribute organic matter to the weathered rock material, and a true soil begins to form.

Since all of these soil-forming agents are in operation constantly, the process of soil formation is continual. Evidence indicates that the soils we depend on today to produce our crops required hundreds and even thousands of years to develop. In this regard, we might consider soil as a non-renewable resource measured in terms of man's life span. Thus, it is very important that we protect our soils from destructive erosive forces and nutrient depletion which can rapidly destroy the product of hundreds of years of nature's work, as well as greatly reduce soil productivity.

SOIL PROFILE

A vertical cross-section through a soil typically represents a layered pattern. This section is called a "profile" and the individual layers are called "horizons". A typical soil profile is illustrated in Figure 1.2.

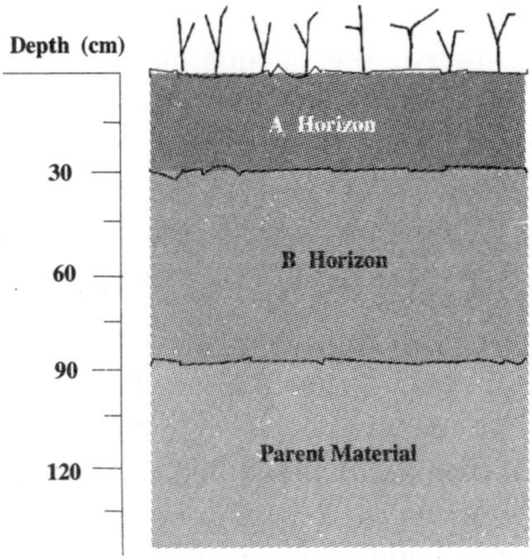

Figure 1.2. A typical soil profile.

The uppermost layer includes the "surface soil" or "topsoil" and is designated as A-horizon. This is the layer which is most subject to climatic and biological influence. It is usually the layer of maximum organic accumulation, has a darker colour, and has less clay than subsoil. The majority of plant roots and most of the soil's fertility are contained in this horizon.

The next successive horizon is called the "subsoil" or B-horizon. It is a layer which commonly accumulates materials that have migrated downward from the surface. Much of the deposition is clay particles, iron and aluminium oxides, calcium carbonate, calcium sulphate and possibly other salts. The accumulation of these materials creates a layer which is normally more compact and has more clay than the surface. This often leads to restricted movement of moisture and reduced crop yields.

The parent material (C-horizon) is least affected by physical, chemical and biological weathering agents. It is very similar in chemical composition to the original material from which the A- and B-horizons were formed. Parent material which has formed in its original position by weathering of bedrock is termed "residual", or called "transported" if it has been moved to a new location by natural forces. This latter type is further characterized on the basis of the kind of natural force responsible for its transportation and deposition. When water is the transporting agent, the parent materials are referred to as "alluvial" (stream deposited). These are often the most productive soils for agricultural crops. Wind-deposited materials are called "aeolian". Climate has a strong influence on soil profile development. Certain characteristics of soils formed in areas of different climates can be described.

The soil profile is an important consideration in terms of plant growth. The depth of the soil, its texture and structure, its chemical nature as well as the soil position on the landscape and slope of the land largely determine crop production potential. The potential productivity is vitally important in determining the level of fertilization.

SOIL TEXTURE

Soils are composed of particles with an infinite variety of sizes. The individual particles are divided by size into the categories of sand, silt and clay. Soil texture refers to the relative proportion of sand, silt and clay in the soil. Textural class is the name given to soil based on the relative amounts of sand, silt, and clay present, as indicated by the textural triangle shown in Figure 1.3. Such divisions are very meaningful in terms of relative plant growth. Many of the important chemical and physical reactions are associated with the surface of the particles, and hence are more active in fine than coarse texture soils.

A textural class description of soils can tell a lot about soil-plant interactions, since the physical and chemical properties of soils are determined largely by texture. In mineral soils, exchange capacity (ability to hold plant nutrient elements) is related closely to the amount and kind of clay in soils. Texture is a major determining factor for water holding capacity. Fine-textured soils (high percentage of silt and clay) hold more water than coarse-textured soils (sandy). Water and air movement through the finer textured soils is reduced and they can be more difficult to work.

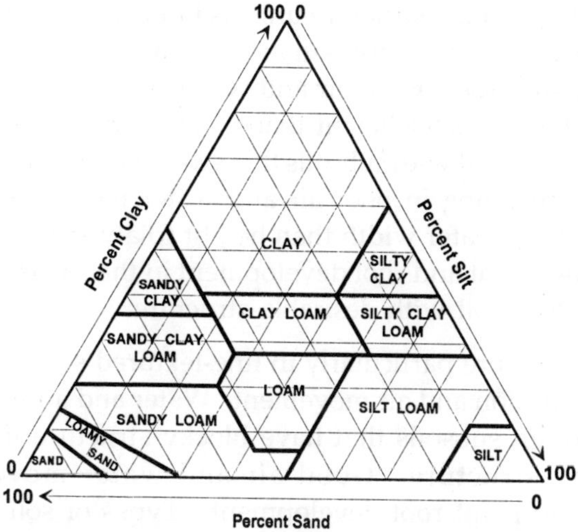

Figure 1.3. Triangle for determining soil textural classes.

From the standpoint of plant growth, medium-textured soils, such as loams, sandy loams and silt loams, are probably the most ideal. Nevertheless, the relationships between soil textural class and soil productivity cannot be generally applied to all soils, since texture is one of the many factors that influence crop production.

Normally, the surface includes the top foot of soil, but it may be shallower or deeper in certain situations. Soil below the tillage zone is called "subsoil". It is also necessary to consider the subsoil texture when determining productivity potentials.

SOIL STRUCTURE

Soil structure refers to the presence of aggregates of soil particles that have been bound together to form distinct shapes. Sometimes the binding or cementing is only weak, however the aggregates are much larger than individual soil particles. Soil organic matter contributes significantly as a cementing agent. Air and water movement and root penetration in the soil is related to the soil structure. The better the structure, the higher the productivity of the soil is.

Size and shape of the structure units is important. When height of the structure unit is approximately equal to its width (blocky structure), we expect good air and water movement. Structure units that have greater height than width (prismatic structure) are often associated with subsoils that swell when wet and shrink when dry, resulting in poor air and water movement. When particles have greater width than height (platy structure) water and air movement and root development in the soil is restricted, compared to a soil with desirable structure.

Granular structure particularly in fine-textured soils is ideal for water penetration and air movement. Water and air move more freely through subsoils that have blocky structure than those with platy structure. Good air and water movement is conducive to plant root development. Types of soil structure are illustrated in Figure 1.4.

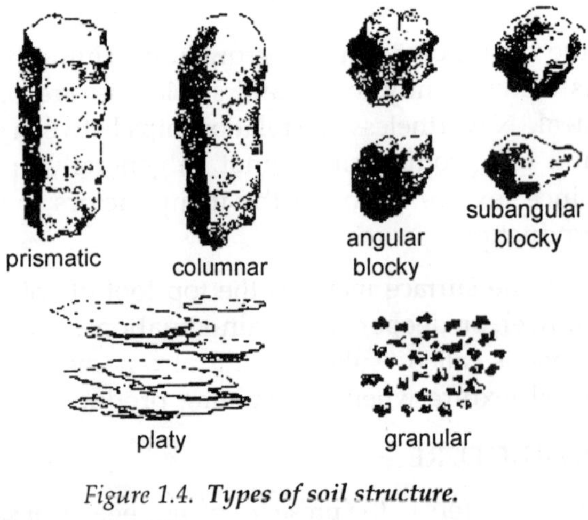

Figure 1.4. ***Types of soil structure.***

The productivity of the soil is influenced by both surface and subsoil texture and structure. An approximate rating for soils considering texture and structure is shown in Table 1.1. Raise or lower the rating 10 to 20 percent, according to whether the soil structure is more or less, favourable than the average. If gravel occurs in the soil, lower the rating according to its effect on the productive capacity.

Table 1.1. Soil productivity rating as affected by texture*.

Subsoil Texture	Surface Soil Texture				
	Sand	Sandy Loam	Loam	Clay Loam	Clay/ Silty Clay
	Percent of maximum productivity				
Sandy	50	55	65	60	55
Sandy Loam	60	70	80	75	65
Loam	70	80	95	90	75
Clay Loam	70	80	90	90	75
Clay/ Silty Clay	65	70	80	80	70

*Numbers represent average soil conditions.

SOIL DEPTH

Soil depth is generally used to describe how deep roots can favourably penetrate. Soils that are deep, well drained, and have desirable texture and structure are suitable for production of most crops. For satisfactory production, most plants require considerable soil depth for root development from which to secure nutrients and water. Plants growing on shallow soils have little soil volume from which to secure water and nutrients. Depth of soil, and its capacity to hold nutrients and water, frequently determines crop yield, particularly for summer crops.

Roots of most crops extend 90 cm or more into favourable soil. Soils should be at least 180 cm deep to give maximum production. Look for materials or conditions that limit soil depth, such as hardpans, shale, coarse gravely layers and tight impervious layers. These are almost impossible to change. On the other hand, a high water table may limit root growth, but it can usually be corrected by drainage. Soil productivity estimates on the basis of soil depth can be made using Table 1.2.

Table 1.2. Soil productivity rating as affected by depth.

Soil depth usable by crop roots	Relative productivity
(cm)	(Percent)
30	35
60	60
90	75
120	85
150	95
180	100

SOIL SLOPE

The earth's surface contour is called its topography. Topography influences soil formation primarily through its associated water and temperature relations. The soil profile on gentle slopes generally is deeper, the vegetation more luxuriant, and the organic matter level higher than in similar materials on steep topography.

Table 1.3. Soil productivity ratings as affected by slope.

Slope	Relative Productivity (%)	
	Stable soil	Unstable, easily eroded soil
0-1	100	95
1-3	90	75
3-5	80	50
5-8	60	30
8-12	40	10

Topography of the land largely determines potential for runoff and erosion, method of irrigation, and management practices needed to conserve soil and water. Higher sloping land requires more management, labour and equipment expenditures.

Table 1.3 can be used to rate land productivity based on slope. If slope varies, steeper slopes can be used for the rating.

EROSION

Principal reasons for soil erosion are: (i) insufficient vegetative cover, which is usually a result of inadequate fertility to support a good plant cover, (ii) growing cultivated crops on soils not suited to cultivation, and (iii) improper tillage of the soil.

Soil erosion can be held to a minimum by (i) using the soil to produce crops for which it is suitable, (ii) using adequate fertilizer and soil amendment to promote vigorous plant growth, and (iii) using proven soil preparation and tillage methods.

Soils that have lost part or all their surfaces are usually harder to till and have lower productivity than non-eroded soils. To compensate for surface soil loss, more fertilization, liming and other management practices should be used.

SOIL AND AVAILABLE WATER

Plants are totally dependent upon water for growth and production. Even with well fertilized soils, limited water can greatly reduce yields. Rainfall is not always dependable, and therefore, crops are dependent upon the moisture stored in the soil profile for growth and production.

Soils differ in their ability to supply water to plants. Limited root zones caused by shallow soils, high water table or clay pans, or extremely porous sub-soils cause drought stress in plants faster than more desirable soils. Differences in available water holding capacity between the soils caused by widely varying textures of the subsoil and soil depth point out the need for knowing what is below the surface. During a drought, differences of 5 cm of available water can keep plants growing for an extra ten days during peak plant use and could be the difference between success and crop failure.

SOIL FERTILITY

Soil fertility is the soil's ability to provide essential plant nutrients in adequate amounts and proper proportions to sustain plant growth. Soil fertility is a component of soil productivity that is quite variable and strongly influenced by management.

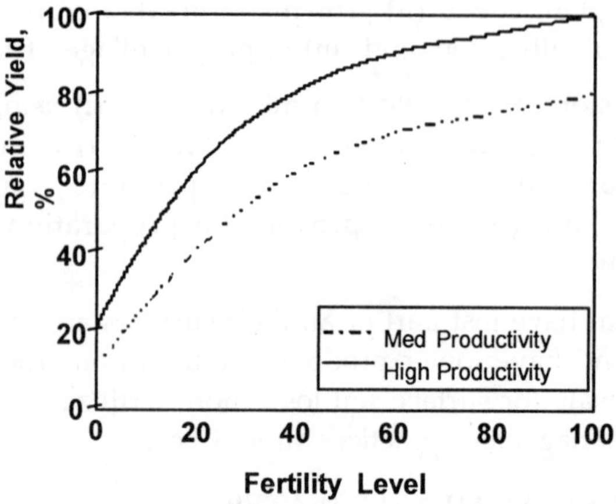

Figure 1.5. ***Influence of soil productivity on yield response to fertility.***

Other components of soil productivity, especially soil slope and soil depth, will be the same year after year. Together with climate, these components set the soil productivity limits, above which yields cannot be obtained even with ideal use of fertilizer. It is important to realize this and understand that added fertilizer cannot compensate for a soil that is unproductive because it is excessively stony or has a subsoil layer that restricts normal root growth and development. This point is illustrated in Fig. 1.5.

SOIL MANAGEMENT

There are numerous other soil characteristics that can be important to soil productivity in specific areas. These include soil drainage, soil salinity and acidity, presence of stone and/or rocks, and organic matter content. They are not major limiting factors over wide areas.

One additional factor on which soil productivity is highly dependent is soil management. This implies using the best available knowledge, techniques, materials, and equipment in crop production. The use of minimum tillage is an important management practice used to reduce the potential damage to soil structure from overworking, and for economic and fuel conservation purposes as well as to allow farming of more acres per unit of labour.

Soil conservation is a concept integrating important management practices which deserves close attention. It is estimated that annually in the India, 5334 million tons of sediment are lost from the land in runoff waters and with it, much of the natural and applied fertility that is equivalent to the total loss of topsoil (six inches deep) from four million acres. Wind erosion is also a problem in certain areas. Management practices such as contouring, strip planting, cover cropping, reduced tillage, terracing and crop residue management help to eliminate or minimize the loss of soil from water and wind erosion.

Proper utilization of crop residues can be a key management practice. Crop residues returned to the soil improve soil productivity through the addition of organic matter and plant nutrients. The organic matter also contributes to an improved physical condition of the soil, which increases water infiltration and storage and aids aeration. This is vital to crop growth.

SUMMARY

Limitations of soil, water, or climate reduce the soil's ability to produce. These limitations increase the need for better management practices. Poor management, or the presence of weeds, compact soils, soil erosion, etc., will result in low yields even on the most productive soils. On the other hand, good management on moderately productive soils can give high yields. Hopefully, by considering the factors discussed in this chapter, one can make a better determination of the soil's overall crop productivity and in turn make better decisions about nutrient management including use of fertilizers.

2

Soil Conditions and Environmental Plant Growth

Plant growth is limited by the soil conditions and environment. If any soil or environmental factor is less than ideal, it will become a limiting factor in plant growth. Limiting factors are also responsible for the geography of plant distribution. For example, only plants adapted to limited amounts of water can live in deserts. Most plant problems are caused by environmental stress, either directly or indirectly. Therefore, it is important to understand the soil conditions and environmental aspects that affect plant growth.

SOIL CONDITIONS

Soils act as reservoirs for the water and nutrients needed by plants. Soils are complex mixtures of mineral particles of various shapes and sizes; living and dead organic materials including microorganisms, roots, and plant and animal residues; air and water. In the soil, physical, chemical and biological reactions occur constantly and are closely interrelated. Optimum plant growth depends as much on a favourable soil conditions as it does on what we call soil fertility.

The physical form of the soil plays a large role in influencing the nature of biological and chemical reactions. The discussion of soil physical characteristics begins with the sizes (texture) and arrangements (structure) of individual soil particles. These two characteristics intimately affect the pore space between the particles. The pore space is important as the conveyor of water, dissolved mineral nutrients and air, as well as for providing space in which roots can grow. Soil colour is discussed because it often provides information about the chemical makeup or status of drainage in the soil. Finally, it is important to consider the whole soil mass, and how it changes (compaction, dryness) with depth below the surface.

Soil Texture

Soil texture is a term which describes the mixture of different sizes of mineral particles. The mineral particles, originally from solid rock, assumed their present form because of physical and chemical processes called weathering. At some stage in the weathering process, mineral particles became a favourable medium for plant growth, that is, they were able to provide storage of water, air, and mineral nutrients, as well as space in which roots could grow. Organic matter then accumulated near the soil surface due to the decomposition of plant residues. Generally, organic matter further improved the properties of the soil as an environment for plant growth.

The percentage of organic matter in soils is quite variable. Well drained soils in the warm, dry regions of world soils often contain less than 1 percent organic matter. Soils in the cool, moist regions contain more than 1 percent but less than 10 percent. Poorly drained valley soils originating in swamps may contain from 5 percent to more than 50 percent organic matter.

Soil texture relates primarily to sand, silt, and clay which are most active in soil processes and support plant growth. Coarser particles, gravel and stones, are either inert or detrimental to plant cultivation. Sandy soils usually have rapid water infiltration and good aeration but low water holding and nutrient storage capacity. Silty soils have a slower water intake rate but a higher water holding capacity than sandy soils. Clay, the finest size fraction largely composed of a different set of minerals, called secondary minerals. These are weathering products of the primary minerals like quartz, feldspar, and mica of which sand and silt are largely composed.

One unusual property of clay is its attraction (called adsorption) for positive ions, such as calcium, magnesium, potassium, ammonium, and others. On the other hand, negative plant nutrient ions such as nitrate, phosphate, and sulphate are repelled by clay particles, and can only be stored for plant use to the extent that they occur dissolved in the water held in soil pores. Clay has a very high affinity for water, partly because of its small particle size and partly because the positive ions associated with clay also attracts water. While soils high in clay are difficult to manage because of their great strength and sticky nature, an intermediate amount of clay in a soil improves its capacity to hold water and plant nutrient ions (either in solution or by adsorption).

Soil Structure

Plant growth is strongly influenced by soil structure. The size and form of aggregation of sand, silt, and clay is known as the structure of the soil. Soil structural aggregate may vary from a fraction of an inch to several inches in diameter; may be approximately spherical, elongated, or plate like; and may be held together strongly or weakly. A granular structure provides an ideal environment for plant roots, and is particularly helpful for establishing plants from seeds or transplants. The larger pores between the granular aggregates are continuous, and roots may penetrate them with ease.

Soil structure affects movement of water, air, and roots through the soil. Water drains readily through this soil, yet moisture is held back sufficiently in the aggregates to supply root needs. Granular structure occurs in loam soils and in some clay soils near the surface. One of the good things about clay is its promotion of granular structure (by swelling and shrinking) in medium textured soils. Greater organic matter content also results in better granular structure of a soil. Sandy soils are low in both organic matter and clay, and aggregation is very weak to nonexistent. The structure is called single-grained; such a soil drains well but does not retain much moisture. Single-grained soils require more frequent irrigation and fertilization for plant roots to thrive.

Prismatic and blocky structures most often occur as the result of shrinking and cracking of clay loams and clay soil layers (called horizons) upon drying. The large cracks that are visible at the surface of dry clay soils may occasionally extend to 90 cm or more in depth. Prismatic or blocky aggregates may vary considerably in size but are always coarser than those of granular structure. The aggregates swell when wet and fit together so tightly that water drains through them rather slowly. Plant roots may follow cracks downward but do not usually penetrate to the centres of prismatic or blocky aggregates. Thus, the roots may not have access to a significant portion of the water and nutrients in these soils.

Platy structure refers to the occurrence of thin layers of soil stacked on top of one another. These most often occur when silty soil materials are deposited in thin layers by stream overflow. The discontinuities caused by this minute layering may interrupt the movement of water, air, and roots into the soil.

If particles are weakly bound together in the whole soil mass, soils are said to have a massive structure. If open and porous, these soils may still provide a favourable root environment. Many massive soils, however, are dense and non-porous providing only slow water and air movement. Compact, massive layers occur naturally in the sub-soils of some old terrace soils, but farming activity has caused similar compaction near the surface of many cultivated soils that originally had granular or single-grained structure. Terrace soils are stream laid soil deposits on longer receiving any deposition of soil materials because of the lowering of stream channels. They occur near the edges of the stream channels. They occur near the edges of the central and coastal valleys, and may be 10,000 to 100,000 years old.

Intensive cultivation usually results in some breakdown of the natural soil structure. Forces holding soil particles together in aggregates may not be strong enough to resist the shearing effect resulting from working the soil at too high a moisture content. Excessive traffic over the land results in a compact soil mass in which large pores have collapsed due to crushing of the granules. In the absence of large pores, water penetration becomes very slow. The small pores, still present, may fill slowly with water after rain/irrigation, and drain even more slowly because water is held strongly by particle surfaces. This has two serious effects. Water movement to lower depths is very slow; and little or no airspace is left in the compacted soil. Feeder roots of most crops die if deprived of air for only a few hours.

The more dense layers resulting from man-made soil compaction usually show up within the surface foot of soil. Regardless of soil permeability beneath the compact layer, water cannot percolate or infiltrate faster than the limiting rate set by the compacted layer. Compaction can develop in almost all soils, although some soils seem more susceptible than others.

Soil Hardness

Hardness of soil strongly affects root growth. The hardness is strongly affected by soil compaction. It is also strongly affected by soil water content, and can vary many folds over the typical range of available water content in the soil. Soil hardness may also affect shoot growth. Such effects are often explicable in terms of the inability of the hampered roots to supply the shoot with water or nutrients.

Soil hardness is usually expressed in terms of penetrometer resistance i.e. the pressure required to push into the soil a cylindrical probe with a conical tip. Root growth slows markedly once this resistance exceeds about 1 MPa and falls away, roughly linearly, to almost nothing at a resistance of about 5 MPa. Figure 2.1 shows the leaf area of plants whose roots were growing in soil of various hardness. The leaf area was closely related to the penetrometer resistance of the soil, even when this was varied by changing the bulk density or the water content of the soil. The elongation of the roots is strongly affected when the penetrometer resistance is large.

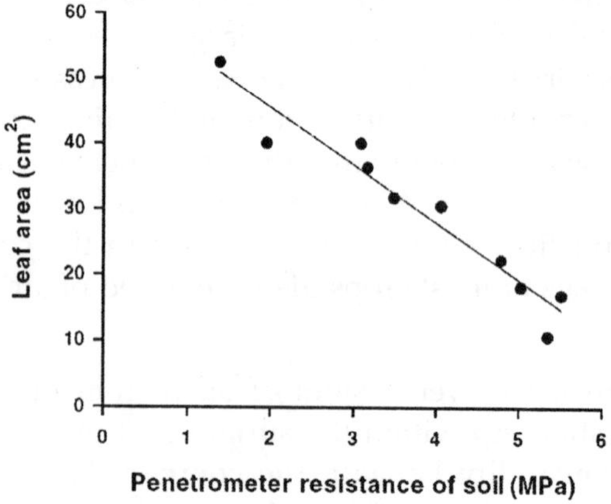

Figure 2.1. Effect of penetrometer resistance of soil on growth of young wheat plants (Source: Masle and Passioura 1987).

Size of Soil Pores

Another physical feature of soil that affects the growth rate of plants without affecting the availability of water and nutrients is that of the size of the pores through which the roots are growing. All plants have adequate supplies of water and nutrients, as with plants growing in hard soil, a signal from the roots affect the growth rate of the shoot. These observations have implications for the performance of plants whose roots are growing in biopores, large continuous pores that have been made and maintained by previous roots or soil fauna. If the biopores are particularly large, for example wormholes, which have diameters many times that of a root, the plants may react adversely even if the roots are able to extract enough water and nutrients.

Figure 2.2 shows evidence that this might be so. Barley plants were grown in cylindrical pots in which soil was packed at a moderate and at a high bulk density. The leaf area was much affected by the roots having access to large pores in the hard soil, being about 30% smaller with 3·2 mm holes than with no holes in the hard soil.

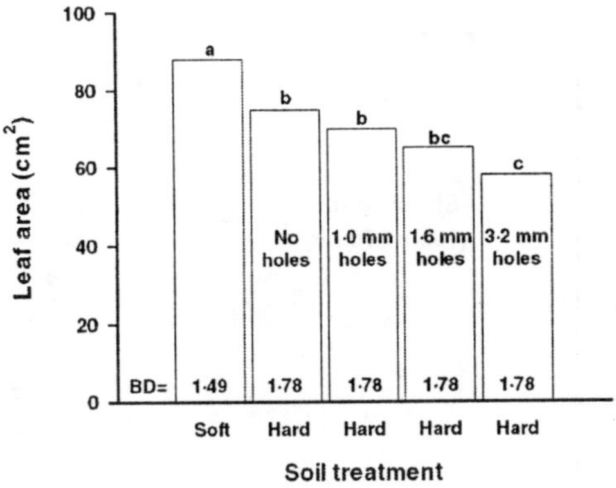

Figure 2.2. Leaf area of young barley plants growing in soil of moderate or high bulk density (BD, Mg/m³) with holes of various diameters formed in the hard soil by pushing steel rods vertically into the soil. Different letters above the bars signify statistically significant (P<0·05) differences among the treatments (source: Passioura and Stirzaker 1993).

Although, such examples show plants growing poorly when their roots are in large pores, such behaviour is overridden if the soil is allowed to dry, for then the ability of the roots in biopores to grow deeply and thereby to access wet subsoil turns the presence of biopores into an advantage. Nevertheless, although biopores may provide highways for the roots, their general sparseness, coupled with the roots' propensity to grow clumped together within them, can drastically slow the uptake of water in comparison with more widely distributed roots.

Excessively Loose Soil

Experience in the field often shows that crops may grow poorly in soil that is too loose. Many such examples can be explained in terms of poor establishment owing to inadequate contact between seed and soil.

Figure 2.3 shows the behaviour of young barley plants growing in soil of very low through moderate to high bulk density. The shoot weights at both very low and very high bulk density were about 15% smaller than those of plants growing at optimal bulk density, whereas the root weight was reduced only at the high bulk density and not at the low.

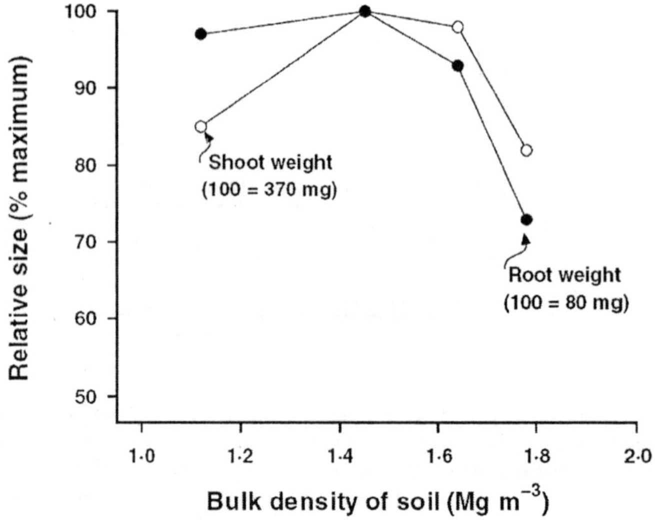

Figure 2.3. Effect of bulk density of soil on shoot and root weight of five-leaved barley plants relative to their maximum values. LSDs ($P<0{\cdot}05$) were 12% (45 mg) for shoot weight and 15% (12 mg) for root weight. (source: Stirzaker et al. 1996).

Soil Drying

When a soil dries, many changes take place within it. It not only holds water more strongly, but it also gets harder and it transmits solutes less readily so that deficiency of poorly mobile nutrients such as phosphorus is more likely. Plants eventually respond to the falling water potential of their leaves by slowing their growth and closing their stomata. However, plants can also react to a drying soil well before there is a detectable change in the water potential of the leaves. Stomatal conductance and the growth rate of the leaves may fall, apparently in response to signals received from the roots in the drying soil. Further, both stomatal conductance and rate of leaf growth may fall as the soil dries even if the water status of the leaves is maintained high by pressurising the roots.

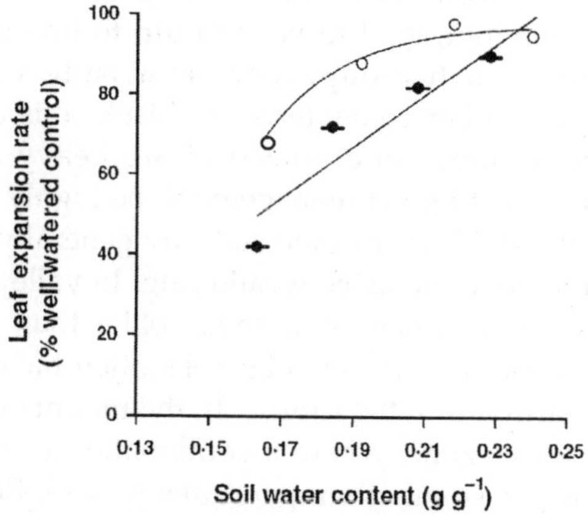

Figure 2.4. Expansion rate of leaves of young wheat plants growing in drying soil of two bulk densities, low, and high. The plants were grown with their roots in pressurized chambers to keep the shoot water potential high at all times. The points marked with a short bar denote that the unwatered plants differed significantly (P<0·05) in leaf expansion rate from the well-watered plants (source: Passioura and Gardner 1990).

Given the common connection between the hardness and the water content of soil, it is conceivable that this early response of plants to soil drying may be as much a response to increasing hardness (Fig. 2.1) as to falling water potential. Figure 2.4 illustrates how this may be so. It shows how the development of leaf area of wheat plants growing in drying soil is affected by both the drying and the hardness of the soil even when the leaf water potential of the leaves is prevented from falling.

Soil Colour

Soil colour is obvious and easily determined. It is one of the most useful characteristics in classification and identification. Although, colour has no direct influence on the functioning or productivity of the soil, a great deal may be inferred about a soil from its colour.

Grey and brown soils are moderately low in organic matter but include some of the most productive alluvial soils. Grey soils are mostly formed from granite alluvium, tend to be coarse to medium textured. The brown soils are formed from sedimentary alluvium, tend to be medium to fine textured. In all areas within each group, there is a wide variation in productivity and other characteristics. Black soils are formed under poorly drained conditions and are heavy (clayey) in texture, but may with good management, be highly productive for field and vegetable crops. Red soils are generally older soils that have undergone intensive weathering. In valleys, red soils occur on terraces or bench lands much older than the soils of the recent alluvial fans. These older soils often have restrictive clay pans or hardpans in the subsoil. In the mountains, red soils occur in the lower zone, where a combination of high winter rainfall and warm summer temperatures prevail. Red soils are often deficient in phosphorous, zinc, and sulphur, in addition to nitrogen. White or light grey soils are usually sandy or calcareous (contain lime). In calcareous soils, iron deficiency may be a problem to some crops, but particularly to orchard crops. Blue of blue-grey layers are usually found in poorly aerated subsoils where organic matter is decomposing anaerobically (without air). These soils contain gases and dissolved materials toxic to plant roots. Extensive aeration is necessary to restore these soils to a condition suitable for plant growth.

Soil Depth

Soil depth is important to the management of plant growth. The deeper the soil, the greater the total water and nutrient storage capacity available to plants.

Root and water penetration through a soil are altered by layers having a distinctly different texture from the layers above or below. If a subsoil layer has a noticeable increase in clay, water may accumulate above this layer, and roots may be injured because of poor aeration. This condition is often called waterlogging.

Very sandy or gravely layers can also interrupt the normal downward penetration of roots or percolation of water. For example, water does not drain freely from a loamy layer into a sandy or gravely layer until the loamy layer becomes saturated for some depth above the coarser layer. When drainage has ceased, a saturated layer remaining just above the textural change will have an adverse effect on roots. The lingering saturated zone remains because particle-to-particle flow of water is poor from the loamy layer into sand or gravel.

Very dense, unfractured rock like layers (hardpan) sometimes occurs in older alluvial soils on relatively flat terraces. These cemented hardpans are impervious to both water and roots. Winter rainfall accumulates above the hardpan but cannot soak through it. Unless the hardpan is shattered and drainage is improved, native grasses or crops grow very poorly on the shallow root zone left as water slowly evaporates from saturated soil.

Many of the soils in the uplands (foothills and mountain areas) rest on hard rock. The density, as well as the degree of fracture of the rock, is quite variable. As a rule, the rock under the soil is more dense in the lower foothills than in the mountainous areas. The density and degree of fracture of the rock are important to moisture storage, drainage, and runoff. A dense, non-fractured, hard rock does not allow water to drain readily from the soil above, nor does the rock store water. A highly fractured rock stores water and allows soil drainage. In a soil underlain with fractured rocks, forest tree roots may extract water to a depth of more than 6 m.

The term effective root depth has been used to describe that portion of the soil favourable for roots. In an alluvial soil, with no noticeable stratification, effective root depth may be more than 1.5 m; in a claypan soil it may be as little as 30 cm, or the depth of soil above the clay layer. Thus, to determine soil depth, it is necessary to determine which layers in the soil will be restrictive to root and water penetration.

Soil Aeration

Oxygen is required by microbe and plants for respiration. Oxygen taken up and carbon dioxide evolved are stoichiometric. Under anaerobic conditions, gaseous carbon compounds other than carbon dioxide are evolved. The manufacturing of sugar by plants requires the presence of carbon dioxide and hence it is one of the vital elements for plant growth. It is a known fact that plants can use as much as 1500 parts per million (ppm) of carbon dioxide.

Oxygen is essential for plant respiration and utilization of photosynthesis byproducts. Root elongation is particularly sensitive to aeration. Oxygen deficiency disturbs metabolic processes in plants, resulting in the accumulation of toxic substances in plants and low uptake of nutrients. Certain plants such as rice are adapted to grow under submerged condition. These have large internal air spaces, which facilitate oxygen transport to the roots.

ENVIRONMENTAL FACTORS

In order for plants to grow, they must intercept light to enable photosynthesis to convert CO_2 into the basic molecules for metabolism such as sugars and amino acids. Water and nutrients are usually acquired from the soil and together with photosynthate, are used to create new plant tissues.

Light

Light has three principal characteristics that affect plant growth i.e. quantity, quality, and duration. Light quantity refers to the intensity or concentration of sunlight and varies with the season of the year. The maximum is present in the summer and the minimum in winter. The more sunlight a plant receives (up to a point), the better capacity it has to produce plant food through photosynthesis. As the sunlight quantity decreases the photosynthetic process decreases.

Light quality refers to the colour or wavelength reaching the plant surface. Sunlight can be broken up by a prism into respective colours of red, orange, yellow, green, blue, indigo, and violet. Red and blue light have the greatest effect on plant growth. Green light is least effective to plants as most plants reflect green light and absorb very little. It is this reflected light that makes them appear green. Blue light is primarily responsible for vegetative growth or leaf growth. Red light when combined with blue light, encourages flowering in plants. Fluorescent or cool-white light is high in the blue range of light quality and is used to encourage leafy growth. These lights are excellent for starting seedlings. Incandescent light is high in the red or orange range but generally produces too much heat to be a valuable light source. Fluorescent "grow" lights have a mixture of red and blue colours that attempts to imitate sunlight as closely as possible. They are costly and generally not of any greater value than regular fluorescent lights.

Light duration or photoperiod refers to the amount of time that a plant is exposed to sunlight. When the concept of photoperiod was first recognized, it was thought that the length of periods of light triggered flowering. The various categories of response were named according to the light length (i.e., short-day and long-day). It was then discovered that it is not the length of the light period but the length of uninterrupted dark periods that is critical to floral development. The ability of many plants to flower is controlled by photoperiod.

Plants can be classified into three categories, depending upon their flowering response to the duration of darkness. These are short-day, long-day, or day-neutral plants. Short-day (long nights) plants form their flowers only when the day length is less than about 12 hours in duration. Long-day (short nights) plants form flowers only when day lengths exceed 12 hours. They include almost all of the summer-flowering plants, as well as many vegetables including beet, radish, lettuce, spinach, and potato. Day-neutral plants form flowers regardless of day length. Some plants do not really fit into any category but may be responsive to combinations of day lengths.

Temperature

Temperature affects the productivity and growth of a plant depending upon whether the plant variety is a warm-season or cool-season crop. If temperatures are high and day length is long, cool-season crops will bolt rather than produce the desired flower. Temperatures that are too low or high for a warm-season crop will prevent fruit set. Temperatures that are too high for warm-season crops such as pepper or tomato can cause pollen to become unviable and not pollinate flowers. Adverse temperatures also cause stunted growth and poor quality. For example, the bitterness in lettuce is caused by high temperatures.

Sometimes temperatures are used in connection with day length to manipulate the flowering of plants. Temperatures alone also influence flowering. Thermoperiod refers to daily temperature change. Plants produce maximum growth when exposed to a day temperature that is about 5.5 to 8°C higher than the night temperature. This allows the plant to photosynthesize and respire during an optimum daytime temperature and to curtail the rate of respiration during a cooler night.

High temperatures cause increased respiration sometimes above the rate of photosynthesis. This means that the products of photosynthesis are being used more rapidly than they are being produced. For growth to occur photosynthesis must be greater than respiration. Low temperatures can result in poor growth. Photosynthesis slows at low temperatures. Since photosynthesis is slowed, growth is slowed and this results in lower yields. Not all plants grow best in the same temperature range.

Plants can be classified as either hardy or non-hardy depending upon their ability to withstand cold temperatures. Winter injury can occur to non-hardy plants if temperatures are too low or if unseasonably low temperatures occur late in the spring or early in the fall. Winter injury may also occur because of desiccation (drying out).

Plant roots need moist soil during the winter. When the soil is frozen, the movement of water into the plant is severely restricted. On a windy winter day broad-leaved evergreens can become water-deficient in a few minutes, turning the leaves or needles brown. Wide variations in winter temperatures can cause premature bud break in some plants and consequent freezing damage. Late spring frost damage can ruin entire crops. If temperatures drop too low during the winter, entire trees of some species are killed by the freezing of plant cells and tissue.

Water (humidity)

As mentioned earlier, water is a primary component of photosynthesis. It maintains the turgor pressure or firmness of tissue and transports nutrients throughout the plant. In maintaining turgor pressure, water is the major constituent of the protoplasm of a cell. By means of turgor pressure and other changes in the cell, water regulates the opening and closing of the stomata, thus regulating transpiration. Water also provides the pressure to move a root through the soil. Among water's most critical roles is that of a solvent for minerals moving into the plant and for carbohydrates moving to their site of use or storage. By its gradual evaporation of water from the surface of the leaf, near the stomata, helps stabilize plant temperature.

Relative humidity is the ratio of water vapour in the air to the amount of water the air could hold at a given temperature and pressure expressed as a percent. Warm air can hold more water vapour than cold air. If the amount of water in the air stays the same and the temperature increases the relative humidity decreases. Water vapour will move from an area of high relative humidity to one of low relative humidity. The greater the difference in humidity the faster water will move. The relative humidity in the air space between the cells within the leaf approaches 100%.

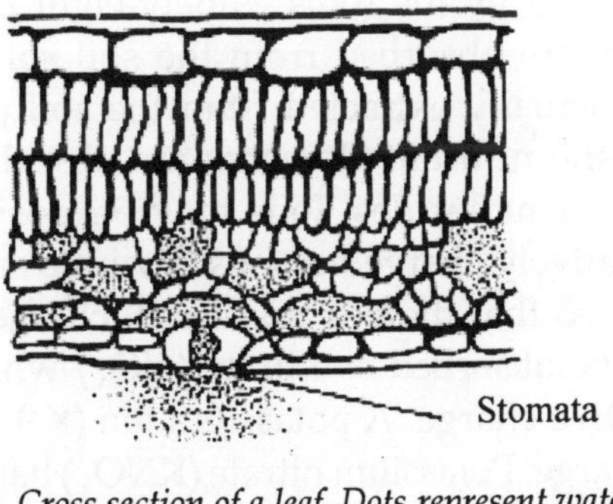

Figure 2.5. Cross section of a leaf. Dots represent water vapour.

When the stomata are open water vapour rushes out (Fig. 2.5). As the vapour moves out, a cloud of high humidity is formed around the stomata. This cloud of humidity helps slow down transpiration and cool the leaf. If air movement blows the humid cloud away, transpiration will increase as the stomata keep opening to balance the humidity.

Nutrition

Plant nutrition refers to the needs and uses of the basic chemical elements in the plant. Plants need 16 elements for normal growth. Carbon, hydrogen, and oxygen are found in air and water. Nitrogen, phosphorus, potassium, magnesium, calcium, and sulphur are found in the soil. The latter six elements are used in relatively large amounts by the plant and are called macronutrients. There are seven other elements that are used in much smaller amounts; these are called micronutrients or trace elements. The micronutrients, which are found in the soil are iron, zinc, molybdenum, manganese, boron, copper, and chlorine. All 16 elements, both macronutrients and micronutrients are essential for plant growth. The macronutrients are either involved in the structure of proteins and nucleic acids, or are key cations involved in charge stabilisation. The micronutrients tend to be in the catalytic sites of specific enzymes or make the prosthetic group of the molecule.

Most of the nutrients that a plant needs are dissolved in water and then absorbed by the roots. Ninety-eight percent of these plant nutrients are absorbed from the soil solution and only about 2% are actually extracted from the soil particles by the root. Most of the nutrient elements are absorbed as charged ions or pieces of molecules. Ions may be positively charged cations or negatively charged anions. Positive and negative are equally paired so that there is no overall charge. For example, nitrogen may be absorbed as nitrate (NO_3^-) which is an anion with one negative charge. A potassium ion (K^+) is a cation with one positive charge. Potassium nitrate (KNO_3^-) has one potassium ion and one nitrate ion. Calcium nitrate ($Ca(NO_3)_2$) has one calcium cation that has two positive charges and two negative, single charge, nitrate ions to match the two positive charges of the calcium.

The balance of ions in the soil is very important. Just as ions having opposite charges attract each other, ions having similar charges compete for chemical interactions and reactions in the environment. Some ions are more active than others or can compete better. For example, both calcium (Ca) and magnesium (Mg) are cations with two charges but magnesium is more active. If both are in competition to be absorbed the magnesium will be absorbed. This explains why the results of a soil test may indicate there is sufficient calcium in the soil, but the plant may still exhibit a calcium deficiency because of an excess of the more active magnesium. What may be expressed as a deficiency in one micronutrient may really be caused by an excess of another.

In order for the ions to be easily absorbed, they must first be dissolved in the soil solution. Some combinations of ions are easily dissolved such as potassium nitrate. When other ions combine, they may precipitate or fall out of solution and thus become unavailable to the plant. Many of the micronutrients form complex combinations with phosphorous and calcium and precipitate out of the soil solution so the nutrients cannot be easily taken up by the plant.

The pH, which means potential hydrogen, is a measurement of acidity or alkalinity. The pH greatly affects chemical reactions. If the soil pH is extremely high (alkaline), many of the micronutrients precipitate out of the solution and are unavailable to the plant. When the soil pH is extremely low (acid), some of the micronutrients become extremely soluble and ion levels may become high enough to injure the plant. The effect of pH varies with the ion, the types of ions in the soil, and the type of soil. Therefore, not only is the amount of the nutrient important but also the soil pH.

Adequate water and oxygen must be available in the soil. Water is required for nutrient movement into and throughout the roots. Oxygen is required in the soil for respiration to occur to produce energy for growth and the movement of mineral ions into the root cells across their membranes. This is an active absorption process utilizing energy from respiration. Oxygen is not transported to roots from the shoot. Without adequate oxygen from the soil environment, there is no energy produced for nutrient absorption. This also stops active absorption in which water flows into the cell due to the higher concentration of nutrients that were actively absorbed.

Anything that lowers or prevents the production of sugars in the leaves can lower nutrient absorption. If the plant is under stress due to low light or extremes in temperature, nutrient deficiency problems may develop. The stage of growth or how actively the plant is growing may also affect the amount of nutrients absorbed. Many plants go into a rest period or dormancy during part of the year. During dormancy, few nutrients are absorbed. Plants may also absorb different nutrients just as flower buds begin to develop.

Nutrients transported from the root to the cell by the vascular system move into the cell through a cell membrane. There are three different ways this happens. First, an entire molecule or ion pair may move through the membrane. If the cell is using energy, called active transport, to absorb ions then only one of the ions in the pair is pulled into the cell. The other will follow to keep the charges even. Most anions (negative ions) are actively absorbed.

The second way of keeping the charges inside the cell balanced and absorbing a new ion is to exchange one charged ion for another ion with the same charge. A hydrogen ion (H^+) is often released so that the cell can absorb another positive ion such as potassium (K^+). Since this is a simple passive exchange, absorption energy may not be required. Cations may be absorbed by this passive method.

Both of the methods mentioned above may be passive or active. The third method, the carrier system, is always active absorption, requiring energy. Scientists have discovered that within the cell membrane there are specialized chemicals that act as carriers. The carrier, through chemical changes, attracts an ion from outside the cell membrane and releases it inside the cell. Once the ion is inside the cell, it is attached to other ions so that it does not move out of the cell. Complex chemical reactions are involved in the entire process.

Although, nutrients can be absorbed passively, research has shown that active absorption must take place if the plant is to grow and be healthy. The factors discussed earlier about absorption by the roots are also true for absorption by the cell. Some of the factors that affect nutrient absorption are the type of ion, soil pH, solubility of ion pairs, water, soil oxygen, sugar supply, plant stress, and temperature.

Under normal growing conditions plants absorb most nutrients, except carbon, hydrogen, and oxygen, from the soil. However, some nutrients can also be absorbed by the leaves if they are sprayed on with a dilute solution. The factors that affect absorption by the cell are still important because the nutrient must enter the cell to be used by the plant.

Nutrient Availability

When individual plants are grown with restricted access to a particular nutrient, growth is reduced and specific deficiency symptoms may develop. A general feature is increased allocation into roots and decreased allocation to leaves and stems. Increased allocation to roots enables the plant to explore greater soil volumes to capture nutrients necessary for continued leaf expansion. Over time, the amount of leaf area or mass per unit root mass is reduced compared to a plant of the same total size given free access to nutrients. Restricted nitrogen or phosphorus availability increases root mass per unit plant mass for monocots, dicots, C4 and woody plants. The responsiveness to nutrient limitation varies between different organs and processes. The greatest effect may be on total plant growth.

In each environment, nutrient concentrations vary between soil horizons, will depend on the soil chemistry and may vary in time depending on temperature and moisture. Different plant species may exploit different niches within the same environment, gaining access to nutrients that are not available to other species. The soil horizon that is explored by roots varies greatly between species. Perennial vegetation can have long-lived roots that penetrate deeply into the soil, whereas ephemeral and annual species tend to colonise surface soil more extensively. Much of the nutrient capital is present in the surface litter and soil layers and so roots proliferate there. The volume of soil that can be explored by a given root mass varies greatly between species and is related to branching patterns and specific root length. Fine roots have a limited lifespan and may not be present in a given patch of soil when a nutrient becomes available. For some species, roots proliferate when they sense the presence of a nutrient.

Some roots have a specialised structure and physiology to enable them to capture a particular nutrient. Having a dimorphic root structure may enable plants to extract nutrients from soil that would otherwise be too dry for roots to function normally. In the deep sands, deep sinker roots can extract water from depth and rehydrate surface soil layers, thereby enabling the finer surface roots to continue to function into early summer. Other roots have developed symbiotic associations with nitrogen fixing bacteria or mycorrhizae to enable them to capture a scarce nutrient. Legumes can fix nitrogen from the air in their nodules in exchange for carbon compounds supplied by the host. Over time, this can lead to a significant build up of organic nitrogen that cycles through the litter and can become available to non-leguminous co-occurring species. Plants that fix nitrogen symbiotically seem to have greater requirements for phosphorus, which is why phosphatic fertilizers are generally applied to improved pastures containing legumes.

Roots of many species form associations with mycorrhizae that improve phosphorus and/or nitrogen nutrition of the host plant. The very fine fungal threads can explore much larger soil volumes for a given amount of biomass than plant roots which enables access to poorly mobile elements like phosphorus. They are also able to chemically extract nutrients bound either to soil particles or in organic matter, which may not be present in the soil solution.

Microorganisms like *Rhizoctonia* root-rot and *Pythium* spp., both of which groups are pathogens, and some *Pseudomonas* spp. which, while not pathogenic, nevertheless inhibit plant growth by releasing toxins. Severe infection with *Rhizoctonia* or *Pythium* undoubtedly affects the ability of the roots to provide the leaves with adequate supplies of water and nutrients. There is a prima facie case that the roots send inhibitors to the leaves. Such inhibitors may originate not in the plant but in the rhizosphere, as could be so with abscisic acid produced by microorganisms.

SUMMARY

Soil consists of mineral particles, organic matter, and pore spaces filled with air or water. The mixture of particle sizes (texture) and their arrangement in aggregates (structure) determine how well plant roots may grow in the soil and obtain from it the necessary water, air, and mineral nutrients. Soil with proper physical conditions, and the right balance of all the minerals and nutrients is one of the essential factors instrumental in plant growth. The type of soil and the quality and the nutrients required in it vary according to the plant species. The right pH balance, which measures the alkalinity or acidity of the soil and presence of certain chemicals, is also instrumental in the growth of plants.

Water is one of the most essential factors required in growth of plants. Water plays a crucial role for efficient photosynthesis, respiration, transpiration and transportation of minerals and other nutrients through the plant. Water is responsible for functioning of the stomatal opening of leaves and also the source of pressure for the directed growth of roots through the soil.

In addition to the soil and water conditions, various environmental factors, which play an instrumental role in enhancing the plant growth light, temperature, humidity and gases like CO_2 and O_2. Adequate light is perhaps one of the most important factors influencing plant growth and it is the quantity, quality and duration of light exposure is what matters. Temperature of the surrounding atmosphere as well as the temperature of the soil matters for the plant growth. Optimum temperature is one of the pre-requisites for many of the plant processes, like photosynthesis, respiration, germination, and flowering. The relative humidity in the air is used by the plants and is crucial for the transpiration of the plants. Transpiration is at its peak cutting hot, windy and dry days while transpiration slows down during cool and humid days.

The manufacturing of sugar by plants requires the presence of carbon dioxide and hence it is one of the vital elements for plant growth. Oxygen is essential for plant respiration and utilization of photosynthesis byproducts.

The availability of nutrients influences plant growth. Nutrient concentrations mainly depend on the soil chemistry, availability of moisture and temperature, and soil microorganisms. Hence, optimum plant growth mainly depends on a favourable soil and environmental conditions.

3

Plant Nutrients and Soil Reactions

More than 100 chemical elements are known to man today. However, only 16 have proven to be essential for plant growth. In order for a nutrient to be classified as essential, certain rigid criteria must be met.

The criteria of essentiality are as follows:

1. The element is essential if a deficiency prevents the plant from completing its vegetative or reproductive cycle.
2. The element is essential if the deficiency in question can be prevented or corrected only by supplying the element.
3. The element is essential if it is directly involved in the nutrition of the plant and is not a result of correcting some microbiological or chemical condition in the soil or culture media.

The essential elements and their chemical symbols are listed in Table 3.1.

Table 3.1. Essential plant nutrients, chemical symbols and sources.

From Air and water (non-mineral)		From soil and/or fertilizers (mineral)			
Element	Symbol	Element	Symbol	Element	Symbol
Carbon	C	Nitrogen	N	Iron	Fe
Hydrogen	H	Phosphorus	P	Manganese	Mn
Oxygen	O	Potassium	K	Zinc	Zn
		Calcium	Ca	Copper	Cu
		Magnesium	Mg	Boron	B
		Sulphur	S	Molybdenum	Mo
				Chlorine	Cl

Three of the 16 essential elements- carbon, hydrogen and oxygen are supplied mostly by air and water. These elements are used in relatively large amounts by plants and are considered to be non-mineral since they are supplied to plants by carbon dioxide and water. The non-mineral elements are not considered fertilizer elements. The other 13 essential elements are mineral elements and must be supplied by the soil and/or fertilizers.

The essential plant nutrients may be grouped into three categories. They are as follows:

1. Primary nutrients - nitrogen, phosphorus and potassium
2. Secondary nutrients - calcium, magnesium and sulphur
3. Micronutrients - iron, manganese, zinc, copper, boron, molybdenum, and chlorine

This grouping separates the elements based on relative amounts required for plant growth, and is not meant to imply any element and is more essential than another.

PRIMARY NON-MINERAL NUTRIENTS

Carbon, Hydrogen, and Oxygen

Carbon is the backbone of all organic molecules in the plant and is the basic building block for growth. After absorption of carbon dioxide (CO_2) by the leaves of the plant, carbon is transformed into carbohydrates by combining with carbon, hydrogen, and oxygen through the process of photosynthesis.

Metabolic processes within the plant transform carbohydrates into amino acids and proteins and other essential components.

PRIMARY MINERAL NUTRIENTS

Nitrogen

Nitrogen (N) is an integral component of amino acids, which are the building blocks for proteins. Proteins in turn are present in the plant as enzymes that are responsible for metabolic reactions in the plant. Because N is so important, plants often respond dramatically to available N.

Soil nitrogen reactions and availability

Most of the N in soils is present as organic nitrogen in the soil organic matter. There are about 1120 kg/ha of N in this form for every 1% organic matter in the soil. However, since the soil organic matter is resistant to further decay, most of this N is unavailable during any given growing season. Normally, each year about 2% of the nitrogen from soil organic matter are released to mineral forms when soils are cultivated. This 22.5 to 45 kg/ha of mineral N is typical of the small amount present in unfertilized soils after cultivation and seed bed preparation.

Nitrogen mineralization and immobilization

Because N release from organic matter is dependent upon decay by microorganisms, which themselves require mineral N, the amount of mineral N available for a crop is in constant flux. Unlike crops, which get their carbon as carbon dioxide from the air, many microorganisms get their carbon by decaying organic matter.

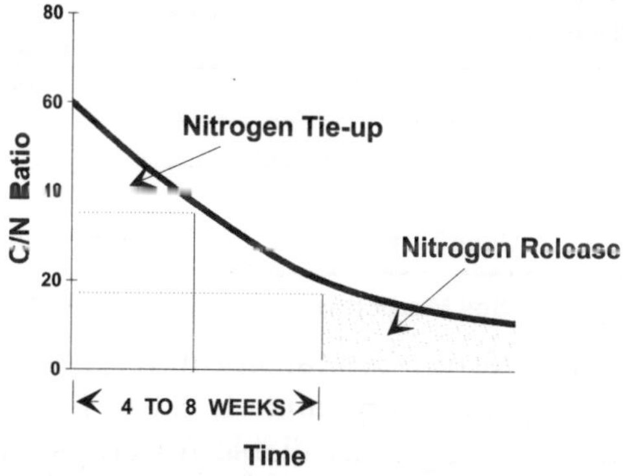

Figure 3.1. Narrowing of carbon to nitrogen ratio as residue is decayed until mineral nitrogen finally becomes available.

Nitrogen availability depends upon the relative amount of carbon and N in the organic matter, its resistance to decay, and environmental conditions to support microbial activity. Figure 3.1 illustrates how nitrogen becomes more concentrated as soil organic matter decays.

Note that nitrogen is not released during the first stages of decay. This is because N that is released is immediately consumed by active microorganisms. With time, remaining organic material becomes more resistant to decay, microorganisms die off, and there is more mineral N present that can be consumed by the few active microorganisms. This results in a final release of measurable mineral N in the form of ammonia (NH_3). The ammonia readily reacts with soil moisture to form ammonium (NH_4^+). These two reactions can be stated simply as:

$$\text{organic N} \rightarrow NH_3 \text{ (gas)} \quad [1]$$

$$NH_3 + H_2O \rightarrow NH_4^+ + OH^- \quad [2]$$

ammonia + water ammonium + hydroxide

Figure 3.2. Interacting pools of soil nitrogen.

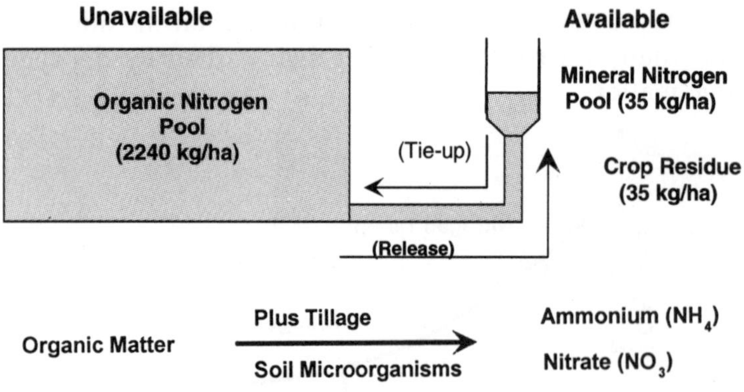

The process of converting or transforming N from organic compounds to inorganic compounds is called mineralization. This results in increasing N available for crops. When the reverse happens and available mineral N is absorbed by crops or microorganisms the process is called immobilization and results in a decrease in the amount of N immediately available for crops. These processes and their interacting nature with soil N for a typical field situation are illustrated in Figure 3.2.

Approximately, 98% of the soil nitrogen is unavailable for plant uptake. This large reservoir of organic N provides an important buffer against rapid changes in available N and plant stress.

The small reservoir of mineral N can often be slowly replenished by mineralization (Fig. 3.2) when crops need additional N.

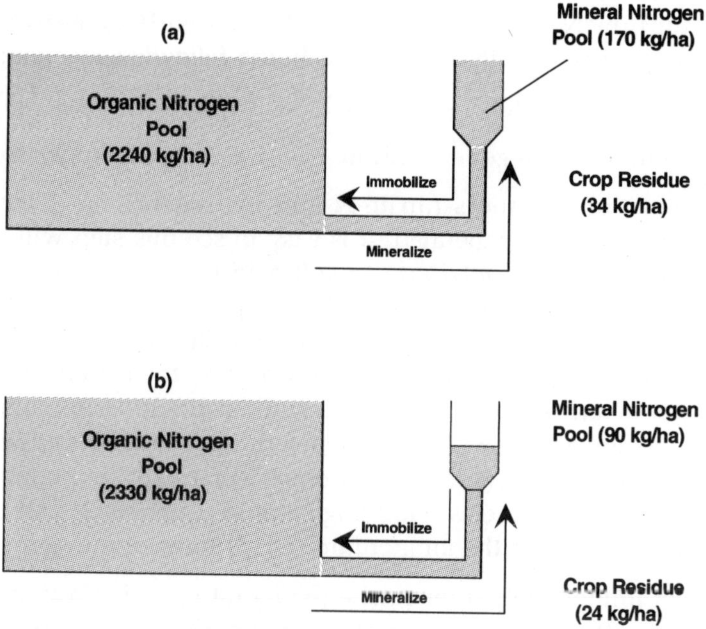

Figure 3.3. *Relative amounts of organic and mineral nitrogen in soil immediately after fertilizing (a) and several days after active immobilization (b).*

Usually, supplemental N as fertilizer must be added to support high and economic production levels. Immediately following fertilization (e.g. with 135 kg N), the system may be illustrated by Figure 3.3a. Addition of fertilizer N will stimulate microorganism activity resulting in consumption of mineral N and breakdown of some crop residues (immobilization) as illustrated in Figure 3.3b. The immobilized N will be present as microbial tissue and other new material in the organic pool. As indicated by the two arrows pointing in opposite pathways, mineralization and immobilization are both taking place simultaneously. Immobilized fertilizer N will again become available in a few weeks if conditions favours crop uptake.

Nitrification

Other reactions, in addition to the general mineralization and immobilization reactions, are responsible for N changes (transformations) in the soil. Nitrification is one of the first reactions to occur after organic N has been converted to ammonium-N. This change is also a result of aerobic microorganism activity as depicted in the following reaction:

$2NH_4^+$	+	$3O_2$ →	$2NO_2^-$ + $2H_2O$ +	$4H^+$ [3]
ammonium	oxygen	nitrite	water	hydrogen ion

This reaction produces nitrite-N and hydrogen ions. Since hydrogen ions are generated, it is easy to see this step will at least temporarily contribute to soil acidity. However, this production of acidity is partially compensated for by the hydroxide (OH^-) produced from reaction [2]. The hydrogen and hydroxide will combine to form water, so the net effect on acidity when organic nitrogen is mineralized. The same reactions and acidity will occur when fertilizer nitrogen is added in the ammonia form (anhydrous ammonia or urea). Ammonium sulphate will be twice as acidifying because equation [2] will be avoided by adding the ammonium (NH_4^+) form of nitrogen.

Almost immediately after nitrite (NO_2^-) nitrogen is produced (reaction [3]), a companion reaction occurs that is also carried out by soil microorganisms resulting in nitrate-N (NO_3-N) being produced from nitrite.

$$2NO_2^- + O_2 \rightarrow 2NO_3^- \qquad [4]$$

Because this change is quite rapid compared to the change from ammonium to nitrite [3] there is seldom any nitrite (NO_2^-) present in soils. Ammonium and nitrate are common and will increase or decrease depending on microbial activity that will both generate and consume ammonium and nitrate. This cyclic interaction of N transformations is shown in Figure 3.4.

Whenever nitrate and/or ammonium nitrogen are measured in the soil, these measurements provide a view of two components of the N cycle at a single point in time. If the measurement is made when the system is likely to be in balance, or equilibrium, then the value can be a useful guide for determining N fertilizer needs. Because ammonium and nitrate nitrogen are the two forms of nitrogen that higher plants utilize, these two forms have received the greatest attention.

When crop production is added to the cycle in Figure 3.4, it becomes obvious that the cycle is not self-sustaining. Harvesting removes significant amounts of nitrogen each year and eventually the system becomes depleted in organic matter and available N to support normal crop yields. A common response to this result is to begin adding nitrogen back by using legumes and commercial fertilizers.

When additions are balanced with removals, soil organic matter and productivity can potentially be sustained. However, excessive tillage, residue removal (straw and chaff in rice / wheat production) and residue burning often result in continued soil organic matter decline. This loss in soil organic matter also contributes to "hard" ground and soil that easily crusts after drying.

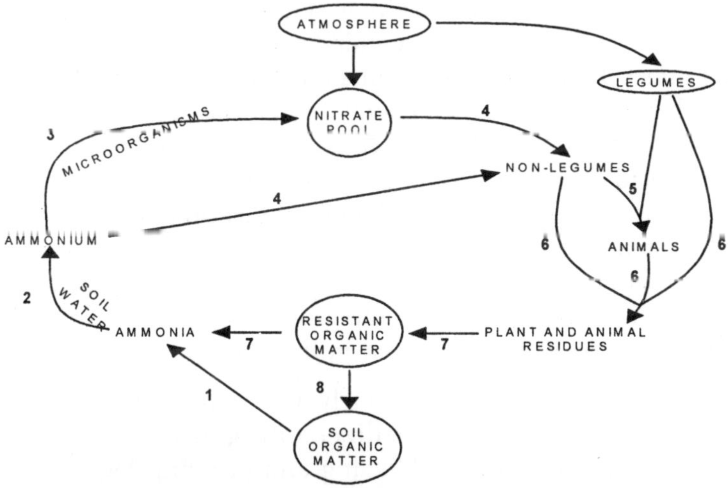

Figure 3.4. ***Primary forms of nitrogen in soils and the transformations among them. (1) Decay of soil organic matter releasing ammonia; (2) reaction of ammonia with water to form ammonium; (3) transformation of ammonium to nitrate by microorganisms; (4) uptake of ammonium and/or nitrate by plants and microorganisms; (5) plants eaten by animals; (6) animal manures, nitrogen fixation and plant residue returned to soil; (7) residues decayed to resistant organic matter, ammonia produced from nitrogen rich materials; (8) soil organic matter produced as decay continues.***

Nitrogen fixation

Additions to soil, N are made as a result of either atmospheric, biological, or industrial fixation of atmospheric nitrogen (N_2). These processes are responsible for transforming N from the atmosphere to either ammonium or nitrate nitrogen that can be used by plants. The atmosphere contains an inexhaustible amount (78%) of nitrogen. Approximately 35,000 tons of N is present in the atmosphere above every acre of the earth's surface.

Atmospheric nitrogen fixation occurs when there is electrical discharge or lightning during thunderstorms. This causes elemental nitrogen (N_2) to combine with elemental oxygen (O_2) to form nitrate (NO_3^-). The nitrate is added to the soil with rainwater and accounts for about 3.3 to 5.6 kg of N per ha per year.

Biological N fixation can be either symbiotic or non-symbiotic. Symbiotic N fixation occurs within legumes. Bacteria (*rhizobium* sp.) infect the root of the legume and cause a nodule to form. The rhizobium obtain their energy from the legume and convert free N to ammonia (NH_3), which the host plant utilizes to make amino acids and proteins. Legumes may fix as much as 550 kgs of nitrogen per ha per year (alfalfa) by this process. However, only a small fraction of the N fixed by legumes will be available for subsequent crops unless the legume is "ploughed down" when a significant amount of top growth is present. Normally, most of the fixed N is removed in the harvest.

Biological N fixation is an extremely important source of adding nitrogen to soils when fertilizer nitrogen is unavailable. However, the cost of establishing and growing legumes for this purpose alone, precludes their use as a sole substitute for nitrogen fertilizers.

Non-symbiotic nitrogen fixation is accomplished by certain "free-living" microorganisms (cyanobacteria or blue-green algae), which live independently of other living tissue. The total contribution of nitrogen from these microorganisms is usually insignificant.

Industrial fixation of nitrogen involves reacting atmospheric nitrogen (N_2) with hydrogen (H), usually in the form of natural gas, under high temperature and pressure to form ammonia (NH_3). The ammonia may be used directly as anhydrous ammonia gas or converted to other nitrogen fertilizers such as urea, ammonium nitrate, urea-ammonium nitrate solution, ammonium sulphate or ammonium phosphates.

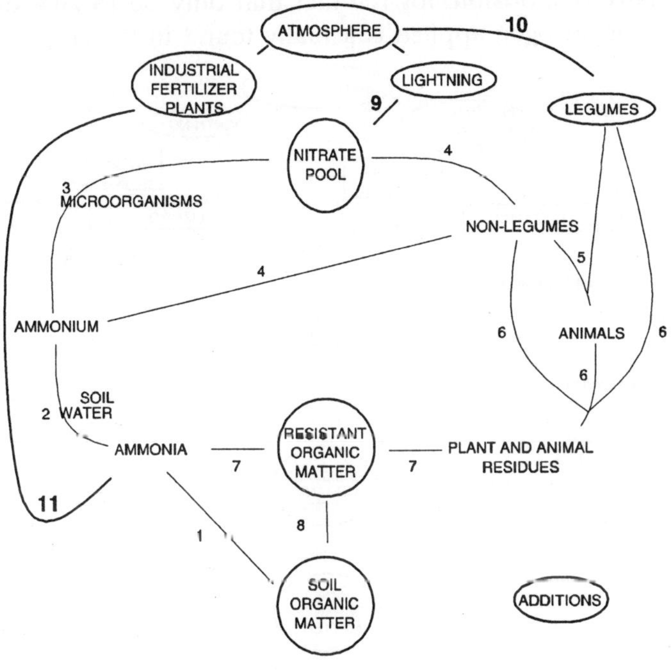

Figure 3.5. *Addition of nitrogen to the nitrogen cycle from fixation of atmospheric N by: (9) lightning; (10) symbiosis with legumes; (11) industrial fertilizer plants.*

Nitrogen fixation results in addition of nitrogen to the soil through utilization by plants and their residues subsequently added back to the soil (Figure 3.5). In order for soil organic matter to be maintained it is necessary for these additions to be at least equal to the amount of nitrogen removed from the field by harvest. Figure 3.5 illustrates how nitrogen fixation interacts with other forms of nitrogen and their transformations.

Each of these above processes is only responsible for very small amounts of nitrogen loss over the course of a crop growing season. However, when considered over a generation of farming, or even just a few years, the amount of nitrogen lost can be significant. Figure 3.6 illustrates the interaction of these nitrogen losses with other forms of nitrogen and their transformations. Nitrogen losses by these processes are at least partially responsible for the fact that only 50 to 70% of the fertilizer nitrogen applied is actually found in the crop.

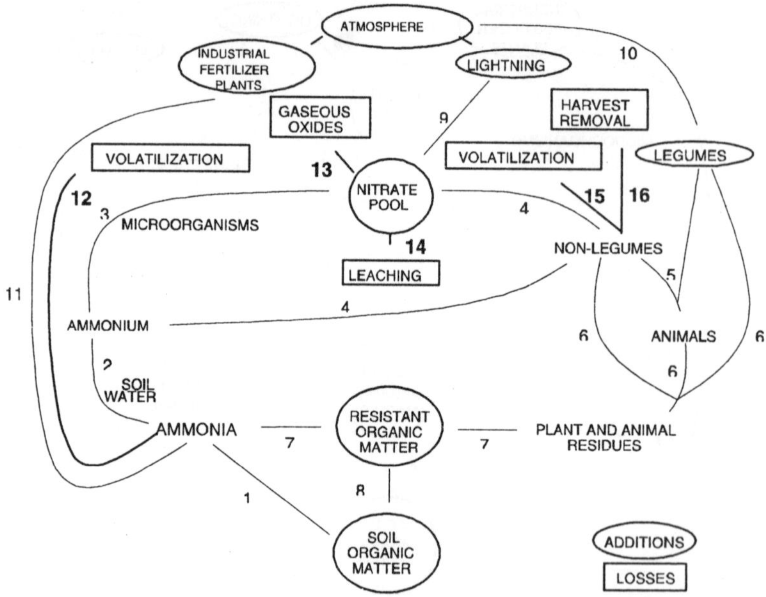

Figure 3.6 ***Losses of nitrogen from the nitrogen cycle as a result of: (12) ammonia volatilization; (13) transformation of nitrate to gaseous oxides (denitrification); (14) leaching below the root zone; (15) volatilization from crops; and (16) harvest removal.***

Phosphorus

Most of the total phosphorus in soils is tied up chemically in compounds with low solubility. In neutral to alkaline pH soils, calcium phosphates are formed, while in acid soils, iron and aluminium phosphates are produced.

Soil phosphorus reactions and availability

Available soil phosphorus, or that fraction which the plant can use, makes up about one percent or less of the total phosphorus in soils. The availability of inorganic phosphorus in soils is related to solubilities of the compounds present. Solubilities are controlled by a number of factors.

The amount of precipitated phosphorus is one factor. The greater the total amount presents in soil, the greater the chance to have more phosphorus in solution. Another important factor is the extent of contact between precipitated phosphorus forms and the soil solution. Greater exposure of phosphate to soil solution and plant roots increases its ability to maintain replacement supplies. During periods of rapid growth, phosphorus in the soil solution may be replaced 10 times or more per day from the precipitated or solid phase. The rate of dissolution and diffusion of soluble P determines soil phosphate availability. As phosphate ions (mainly $H_2PO_4^-$ and HPO_4^{2-}) are taken up by the plant, more must come from the solid phase.

Soil pH can be a controlling factor that determines phosphorus solubility. Maximum phosphorus availability occurs in a pH range of 5.5 to 7.2. At soil pH levels below 5.5, iron (Fe), aluminium (Al) and manganese (Mn) react with phosphorus to form insoluble compounds. When soil pH exceeds 7.2, phosphorus will complex with calcium (Ca) to form plant unavailable P forms. However, it should be noted that the solubility of calcium phosphates is much greater than aluminium and iron phosphates.

The proportion of total soil phosphorus that is relatively available is dependent upon time of reaction, type of clay present in the soil, organic matter content and temperature. The solubility of phosphate compounds formed from added phosphorus due to time of reaction can be broken down in three major groups (Fig. 3.7).

Very slowly available phosphates
apatite, aged Fe, Mn and Al phosphates,
stable organic phosphates

↓↑

Slowly available phosphates
$Ca_3(PO_4)_2$, freshly formed Fe, Al, Mn phosphates
(small crystals), and mineralized organic phosphates

↓↑

Readily available phosphates
water-soluble ammonium phosphates
$NH_4H_2PO_4$ (MAP 11-52-0)
$(NH_4)_2HPO_4$ (DAP 18-46-0)
monocalcium phosphate
$Ca(H_2PO_4)_2$ (0-46-0)
water-insoluble dicalcium phosphate
$CaHPO_4$

Figure 3.7 ***Relative availability of different phosphate forms and their transformations.***

Fertilizer phosphates are generally in the readily available phosphate group but are quickly converted to slowly available forms. These can be utilized by plants at first, but upon aging are rendered less available and are then classified as being very slowly available. At any one time, 80 to 90 percent of the soil phosphorus is in very slowly available forms. Most of the remainder is in the slowly available form since less than 1 percent would be expected to be readily available.

The formation of insoluble phosphorus containing compounds in soils that renders phosphorus unavailable for plant use is called *phosphorus fixation.* Each soil has an inherent fixation capacity that must be satisfied in order to build available phosphorus levels. It is important to understand that the actual amount of P in the soil and the amount available to crops will not necessarily be reflected in a soil test. These soil tests simply provide an index of sufficiency and not an index of build-up or accumulation. Because different soils will have differing fixation capacities, the importance of annual soil testing becomes clear since this practice is the only method we have to estimate future crop fertilizer needs. In addition, these tests should reflect past management (farmers applying more than enough or not enough on an annual basis) and thus compensate accordingly.

Organic matter and microbial activity affect available soil phosphorus levels. As was the case with nitrogen, the rapid decomposition of organic matter and consequent high microbial population results in temporary tying up of inorganic phosphorus (immobilization) in microbial tissue, which is later rendered available through release (mineralization) processes. This is one of the reasons why broadcasting P in zero/minimum tillage systems can be beneficial, especially where soil P fixation capacities are high.

Less than 30% of the phosphorus fertilizers applied is actually recovered in the plants. Therefore, due to fixation reactions, more P must be added than is actually removed by crops. Legumes, in general, require much larger amounts of P than many of the common non-legume crops.

Because phosphorus is immobile in the soil, roots must come in direct contact with this element before the plant can take it up. However, phosphorus is mobile within the plant and if deficient, lower leaves will generally demonstrate outer edge leaf margins that are purple in colour.

Over a wide range of soils and cropping conditions, phosphorus has proven to be one of the more deficient elements in Indian agriculture production. Soil testing on an annual basis should assist in determining crop needs.

Potassium

Plants take up potassium as the potassium ion (K^+). Potassium within plants is not synthesized into compounds and tends to remain in ionic form in cells and plant tissue. Potassium is essential for photosynthesis, starch formation and translocation of sugars within plants. It is necessary for the development of chlorophyll, although it is not part of its molecular structure.

Soil potassium reactions and availability

In most soils (except those that are extremely sandy in high rainfall regions), total potassium contents are high. Similar to nitrogen and phosphorus, not all of the total potassium is available for plant growth. The relationship of unavailable, slowly available and readily available forms of potassium is illustrated in Figure 3.8.

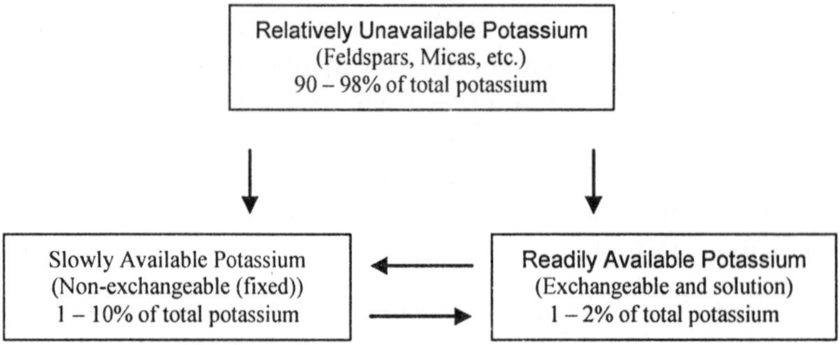

Figure 3.8. ***Relative amounts of soil potassium present in different levels of availability to plants.***

Only 1 to 2 percent of the total potassium in soils is readily available. Of this, approximately 90 percent is exchangeable or attached to the outside edge of clays and the remaining 10 percent is in the soil solution. Equilibrium exists between the non-exchangeable, exchangeable and water soluble forms. When the plant removes potassium from the water soluble form, the concentration is readjusted by the exchangeable and non-exchangeable forms. In the case of added potassium, some of the available forms will move toward non-exchangeable forms. The non-exchangeable form may also be referred to as fixed. Certain 2:1 type clay minerals have pore space large enough for the potassium ions (K^+) to become trapped, rendering the ions unavailable for immediate plant use. Potassium is positively charged and clays are negatively charged and this makes the potassium ion relatively immobile in the soil. Except in extremely sandy soils, leaching losses are minimal. The largest loss comes from crop removal, particularly where hay crops are produced.

Since K is mobile in the plant, visual deficiency symptoms usually appear first in the lower leaves, and progress to the top as the severity of the deficiency increases. Necrotic lesions on broadleaf plants, chlorotic and necrotic leaf margins on grasses, straw lodging in small grains, and stalk breakage in corn.

Interactions with other nutrients. K^+ enhances NH_4^+, NO_3^- and Cu^{2+} uptake; reduces B and Mo uptake and decreases Ca^{2+} and Mg^{2+} in plant tissue. K^+ also enhances Mn^{2+} uptake when Mn is deficient and decreases uptake when Mn is present in toxic amounts. High NH_4^+ with inadequate K^+ may cause toxicity symptoms. K^+ reduces Fe^{2+} toxicity.

SECONDARY MINERAL ELEMENTS

Nutrients that are used in relatively moderate amounts by most plants have been categorized as secondary nutrients. These nutrients are calcium (Ca), magnesium (Mg), and sulphur (S).

Calcium

Calcium is taken up by plants as the cation, Ca^{2+}. Calcium functions in the plant in cell wall development and formation. Calcium is not translocated in plants and consequently, the deficiency of calcium is observed first in the new, developing plant tissue. Calcium deficient tissue fails to develop normal morphological features and is appear to be an undifferentiated gelatinous mass in the region of new leaf development.

The calcium ion is referred to as a basic ion because the element reacts with water to form the strong base calcium hydroxide, $Ca(OH)_2$. Calcium is held tightly on the negatively charged clay and organic particles in soils and is abundant in soils that have developed in arid and semi-arid climates. Because of this, it is primarily responsible for maintaining these soils at or near a neutral pH. In addition to unweathered primary and secondary minerals, soils often contain calcium in the form of impure lime (calcium carbonate, $CaCO_3$) and gypsum (calcium sulphate, $CaSO_4$).

Interactions with other nutrients. Since Ca^{+2} is so directly related to pH in solution, it affects all of the other nutrients.

When NO_3-N is applied to soil, Ca^{+2} absorption increases in the plant. Increases in Ca^{+2} in soil decreases Al^{+3} in acid soils, as well as decreasing Na^{+} in sodic soils. Increases in Ca^{+2} taken up by plants cause deficiencies of Mg^{+2} and K^{+}. MoO_4^{-2} and $H_2PO_4^{-}$ availability increases with increases in Ca^{+2} concentrations.

For most soils, before the available calcium level reaches a critically low point, the soil pH become so low that soil acidity may be a major limitation to crop production. Since the common correction of acid soils is to add lime in amounts of tons per hectare, this practice incidentally maintain a high level of available calcium for crops.

Magnesium

Magnesium is absorbed as the divalent cation, Mg^{2+}, and functions in many enzymatic reactions as a co-factor or in a co-enzyme. The most noteworthy function of magnesium in plants is as the central cation in the chlorophyll molecule. Without magnesium, plants cannot produce adequate chlorophyll and will lose their green colour and ability to carry out photosynthesis, the process responsible for capturing energy from sunlight and converting it into chemical energy within the plant. Magnesium deficiency result in yellow, stunted plants.

Magnesium reactions in soils are similar to calcium in many respects. Magnesium, like calcium is a basic ion that is generally abundant in arid and semi-arid soils that are near neutral in pH. Uptake of K^{+}, NH_4^{+}, Ca^{2+}, Mn^{2+} by plant limits Mg^{2+} uptake; H^{+} (low pH) can limit Mg^{2+} uptake and Mg salts increase phosphorus adsorption.

Like calcium, deficiencies are likely to occur on acid soils, and since most lime contain a small amount (2 to 5%) of magnesium carbonate, liming acid soils on a regular basis usually assure an abundant supply of plant available magnesium. If magnesium deficiency is a reoccurring problem, dolomitic lime (primarily magnesium carbonate) should be sought as a liming source. Deficiencies most often occur in deep sandy soils with a history of high forage production (20 to 25 tons per ha annually), where forage has been removed as hay.

Sulphur

Sulphur is absorbed by plants as the sulphate anion, SO_4^{2-}. Sulphur is a component of three of the 21 essential amino acids and thus, is critical to the formation and function of proteins. Sulphur deficiency causes plants to become light green and stunted. Most crops require about $1/20^{th}$ amount of sulphur that they do of nitrogen. Bumper yields of most crops can be supported by 5 to 15 kg/ha of sulphur.

Sulphur is found in soil in the form of soil organic matter (like nitrogen), dissolved in the soil solution as the sulphate ion, and as a part of the solid mineral matter of soils. Sulphur compounds, like gypsum for example, are slightly soluble in water. Like nitrate nitrogen, the negatively charged sulphate ion is not readily adsorbed to clay and humus particles and may be leached into the subsoil with a porous surface soil layer. Sulphur associated with salts and exchangeable cations, can be replaced by phosphorus on exchange sites.

Sulphur deficiencies most often occur in deep sandy soils, low in organic matter, with a history of high crop production and removal. Soils that have a well developed B horizon seldom will be deficient in sulphur because sulphur will not leach out of the root zone and the accumulated sulphur in the subsoil will adequately satisfy crop needs. This is one of the reasons why early S deficiencies often disappear at late stages of growth, at which time roots have penetrated subsoil horizons rich in sulphur. Soils that contain normal amounts of organic matter releases sulphur by mineralization, much like nitrogen, and this contribute significantly to meeting crop needs. Sulphur is still added incidentally as a component of phosphate fertilizers and other agricultural chemicals which contribute significantly to the requirement of crops.

Micronutrients

The micronutrients are grouped together because they are all required by plants in very small amounts. Some, like molybdenum (Mo), are required in such small amounts that deficiencies can be corrected by applying the element at only a fraction of a kg per hectare. Similarly, chlorine is needed in such small quantities that when researchers at the University of California were attempting to document its essentiality, they found that touching plant leaves with their fingers transferred enough chlorine from the respiration on their skin to meet the plant's requirements. These elements do not function in plants as a component of structural tissues like primary and secondary nutrients. Instead, micronutrients are mainly involved in metabolic reactions as a part of enzymes where they are used over and over without being consumed. Nevertheless, their functions are very specific and cannot be substituted for by some other element. Deficiencies of any of the elements can be corrected by foliar application of solutions containing the element.

Iron

Iron is a component of cells, proteins, and enzymes. It is involved in nitrogen fixation, respiration and photosynthesis. Fe^{+3} (Ferric) is reduced to Fe^{+2} at the root surface before it is absorbed. Soil $Fe(OH)_3$ is usually the most soluble form of iron in soils and, therefore, typically controls the solubility of iron in aerobic soils.

Soil Iron Reactions and Availability

Soils contain about 1 to 5% iron, which is many fold more than that required for plants, however, in aerobic environments Fe is mainly present in the Fe^{+++} oxidation state as iron oxide (written as either $Fe_2O_3 \cdot nH_2O$ or $Fe(OH)_3$) which is very insoluble. The amount of Fe^{+++} in aqueous solution is governed by

$$Fe(OH)_3 \leftrightarrow Fe^{+++} + 3(OH)$$

for which the equilibrium condition is expressed in molar concentrations as

$$(Fe^{+++})(OH)_3 \ / \ Fe(OH)_3 = 10^{-39.4} \qquad [5]$$

Since $Fe(OH)_3$ is a solid, it has an activity of unity [5] and the equation becomes

$$(Fe^{+++})\ (OH)^3 = 10^{-39.4} \qquad [6]$$

and the value $10^{-39.4}$, instead of being called the equilibrium constant (K_{eq}), is called the solubility product constant (K_{sp}). The concentration of Fe^{+++} in solution is given by

$$Fe^{+++} = 10^{-39.4} / (OH)^3 \qquad [7]$$

and at pH = 7.0,

$Fe^{+++} = 10^{-39.4}/(10^{-7})^3$; $=10^{-39.4} / 10^{-21}$; $= 10^{-18.4}$ moles/litre.

Since the atomic weight of Fe is 55.85, the concentration in ppm would be

55.85 g/mole x 1000 mg/g x $10^{-18.4}$ moles/litre

= 55.85 x $10^{-15.4}$ mg/litre; = 55.85 x $10^{-15.4}$ ppm

Fe functions as a co-enzyme, an important electron transfer enzymes, and the formation and component of enzymes that are precursors to chlorophyll.

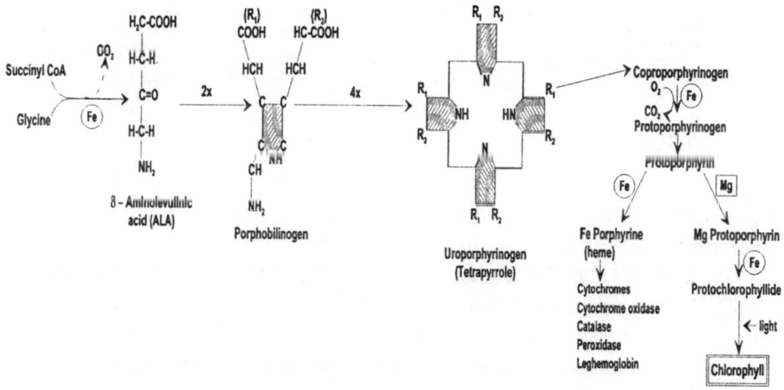

Figure 3.9. Role of Fe in biosynthesis of heme coenzymes and chlorophyll.

The two important categories of enzymes are the Fe-S proteins and the heme proteins. Heme proteins are characterized by a tetrapyrrole ring structure that has Fe as the centrally coordinated metal. Fe is involved as a co-factor in the synthesis of protoporphyrin, which is the precursor to both heme and chlorophyll as depicted in Fig. 3.9. The Fe-S proteins are formed when Fe is coordinated to the thiol group of cysteine, or inorganic S, or both.

The best known Fe-S protein is ferredoxin, important in both reductase and nitrogenase. Other Fe-S proteins have important functions in the citric acid cycle, respiration, SO_4 and SO_3 reduction, and chlorophyll (Fig. 3.10).

Because Fe is strongly bound it is not easily translocated and should be considered immobile in plants. The characteristic deficiency symptoms are interveinal chlorosis in the new leaves of growing plants.

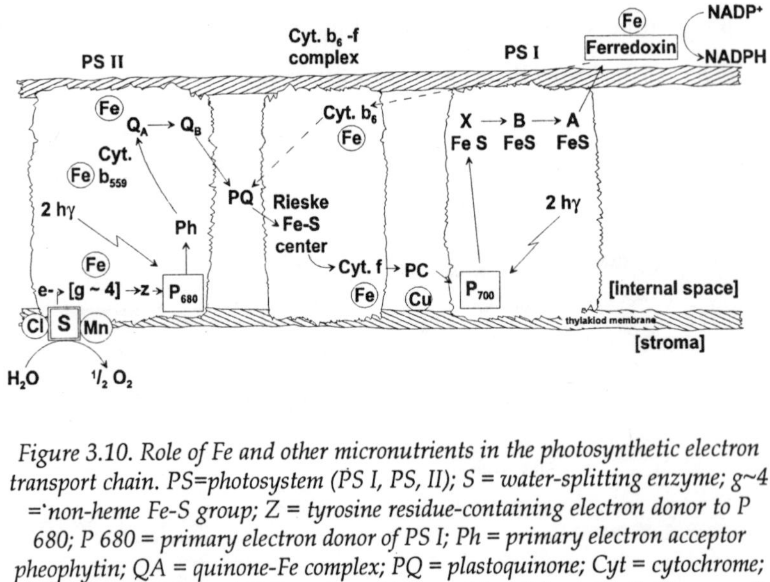

Figure 3.10. Role of Fe and other micronutrients in the photosynthetic electron transport chain. PS=photosystem (PS I, PS, II); S = water-splitting enzyme; g~4 ='non-heme Fe-S group; Z = tyrosine residue-containing electron donor to P 680; P 680 = primary electron donor of PS I; Ph = primary electron acceptor pheophytin; QA = quinone-Fe complex; PQ = plastoquinone; Cyt = cytochrome; PC = plastocyanin; and X, B, and A = Fe_4S_4 proteins. Schematically drawn as Z scheme.

The deficiency concentration of Fe in mature plant tissue is about 50 ppm. Total Fe may be much higher than this level, even in Fe-chlorotic plants because Fe in the plant is not always all metabolically active. HCl extractable Fe is sometimes assumed to be metabolically active and a better guide to plant sufficiency. Fe in plants is found in the Fe^{+++} state, any Fe^{++} is present only as a transitory state (free Fe^{++} is phytotoxic).

The plant's dilemma. The concentration of Fe necessary to provide plants a sufficient amount of Fe by passive uptake has been suggested to be about 10^{-6} moles/litre. At pH 7 the soil supply, as identified by equation [7] is 12 orders of magnitude too small. Even at pH 5 the difference between supply and requirement is still 6 orders of magnitude too small.

Two things are obvious; (a) the plant cannot get enough Fe by passive uptake from the soil solution, and (b) there will be a 1000 fold decrease in supply of available Fe from $Fe(OH)_3$ in the soil with each unit increase in soil pH. Consequently, Fe deficiency to be most common in high pH soils and least in acid soils. This is in fact what is observed. But, how do plants get enough Fe, and why are not all plants subject to Fe chlorosis when grown in neutral and alkaline soils?

Part of the solution to the plant's dilemma of getting enough Fe from the soil is found in chemical reactions called metal chelation. This is the process whereby metals are bound in ring-like structures of organic compounds. The more rings in the structure that the metal is a part of, the stronger the metal is bound. The chemical forces involved are mainly coordinate bonds where the metal acts as a Lewis acid (electron acceptor) and the chelating material has functional groups (sometimes called ligands), like amino, hydroxyl, and phenolic groups that act as Lewis bases (electron donor). The transition metals seek to fill the *d* orbital to attain the electron configuration of the inert gas of that period, krypton. Heme and chlorophyll are examples of natural chelates that hold Fe and Mg as a centrally coordinated atom.

As an example of chelates, consider the common synthetic chelate EDTA. EDTA stands for the chemical compound ethylenediaminetetraacetic acid.

Note that in the Fe-EDTA complex, there are five rings formed with Fe, and that the complex has a single negative charge. As a result, the complex is mobile in the soil and so is the Fe it is carrying. Two other common synthetic chelates are DTPA (commonly used in micronutrient metal soil test extraction procedures) and EDDHA (a commercial chelate for supplying Fe in calcareous soils).

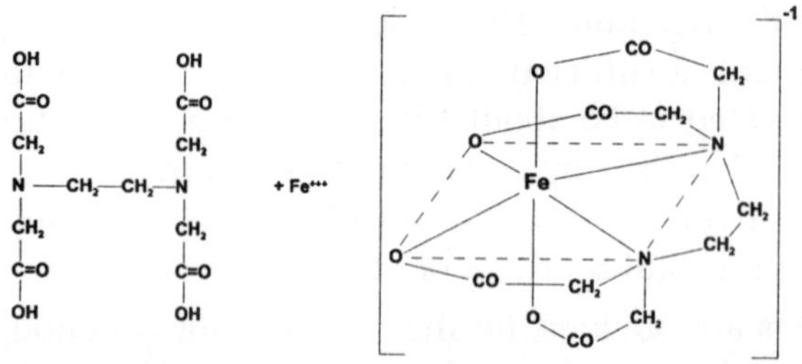

Two important natural chelating compounds in plants are citrate and hydroxamate. Citrate is important as a carrier for the micronutrient metals Cu, Zn, Fe, and Mn. Hydroxamate is a siderophore (produced by micoorganisms) believed responsible for complexing Fe in calcareous soils and increasing its availability to plants.

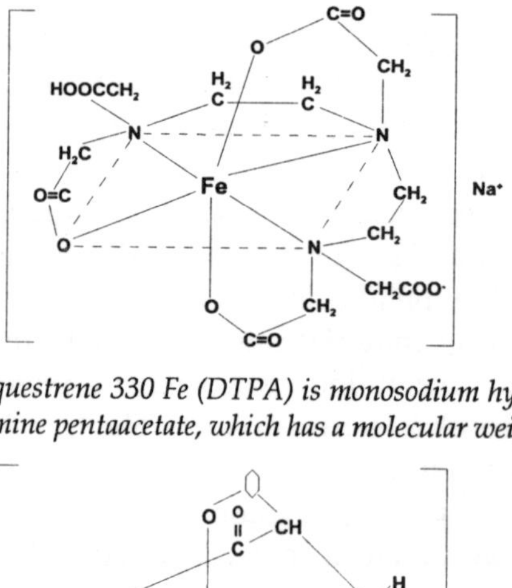

Figure 3.11. *Sequestrene 330 Fe (DTPA) is monosodium hydrogen ferric diethylenetriamine pentaacetate, which has a molecular weight of 468.*

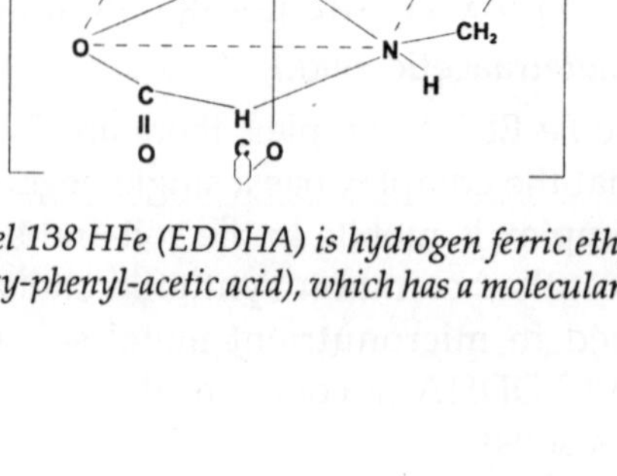

Figure 3.12. *Chel 138 HFe (EDDHA) is hydrogen ferric ethylene bis (alpha-imino-2-hydroxy-phenyl-acetic acid), which has a molecular weight of 413.*

The reaction of chelates with metals to form soluble metal chelates is given by the general equilibrium reaction

$$M + L = ML \qquad [8]$$

Where M refers to the metal concentration, L the chelate (or Ligands) concentration, and ML the concentration of metal chelate. At equilibrium the relative amounts of each present are described in relation to the equilibrium constant as

$$ML / (M)(L) = K_{eq} \qquad [9]$$

Since the equilibrium condition strongly favours the reaction to the right (equation [8]), K_{eq} is called the formation constant K_f. The formation constants for citrate and hydroxamate are $10^{12.2}$ and 10^{32}, respectively.

The benefit of chelates for improving Fe availability can be demonstrated by considering just the reactions involved in chelates complexing Fe from $Fe(OH)_3$. If the reactions are considered simultaneously they can be written as follows for an equilibrium situation where a weak chelate such as citrate is present.

			Log_{10} K	
$Fe(OH)_3$	$\longrightarrow$	$Fe^{+++} + 3(OH^-)$	-39.4	
$Fe^{+++} + L$	$\longrightarrow$	FeL	12.2	
$3(OH) + 3H^+$	$\longrightarrow$	$3H_2O$	42	
$Fe(OH)_3 + L + 3H^+$	$\longrightarrow$	FeL	14.8	
$Fe(OH)_3 + L + 3H^+$	$\longrightarrow$	FeL	$10^{14.8}$	[10]

Equation [10] was obtained by summing the equations (cancelling components that appear as both reactant and product of the reactions) and the $\log_{10}$ of the solubility and formation constants.

The concentration of products and reactants can be expressed for the general reaction in terms of the equilibrium constant as

$$(FeL) / (L)(H^+)^3 = 10^{14.8}$$

or in terms of (FeL) as

$$(FeL) = 10^{14.8} \times (L)(H^+)^3$$

If the soil pH is 7 and the concentration of citrate is 10^{-6}, then the concentration of FeL is

$$(FeL) = 10^{14.8} \times (10^{-6})\,(10^{-7})^3$$

$$(FeL) = 10^{-12.2}$$

Compared to the concentration of Fe^{+++} in solution from $Fe(OH)_3$ dissolving, which is $10^{-18.4}$, this is an improvement of $10^{6.2}$ ($10^{-12.2}$ / $10^{-18.4}$). In other words, the presence of even a weak chelating agent like citrate improve the availability of iron a million fold.

In the case of a metal nutrient like Fe, the concentration of FeL in the soil solution is mainly a function of the formation constant (K_f) and the concentration of L since the other factors are constant.

For example, consider Eq. [9]

$$ML / (M)\,(L) = K_{eq}$$

This can be rewritten as

$$ML = K_{eq}\ (M)\,(L)$$

Where M is Fe^{+++}, and is a constant identified by the solution pH and K_{sp} for $Fe(OH)_3$. Since K_{eq} is also a constant, these can be combined into one constant, to give

$$ML = K\,(L) \qquad [11]$$

Equation [11] identifies that any condition that results in increasing the concentration of L for complexing or chelating Fe will increase the concentration of FeL and thus the availability of Fe for the plant. The two most obvious ways of increasing L will be by (i) drying the soil so the water soluble L will become more concentrated and, (ii) producing more L in the soil solution.

In fact, a common observation is that Fe chlorosis is lessened in susceptible plants when there is definite drying cycles as opposed to continuously moist soil. Also, Fe chlorosis can often be lessened by incorporating large amounts of decayed or decaying organic matter to the soil which will directly provide more L.

Plant absorption of soil Fe (Dicotyledons). Until recently, the mechanisms responsible for allowing some species of plants to grow well in calcareous soils while others commonly exhibited iron chlorosis were not well understood. The observation that some dicots were "iron efficient" while others, even varieties within a species, were "iron inefficient" is explained by an adaptive response mechanism inherent in iron efficient plants. Characteristics of this mechanism, which is activated when plants are in a Fe stress situation, are-

(1) Enhanced root-associated Fe^{+++} reduction

(2) Enhanced H^+ efflux from roots

(3) Accumulation of citrate in roots

(4) Increased root hair development

(5) Increased absorption of Fe

A description of the mechanism is illustrated in Fig. 3.13.

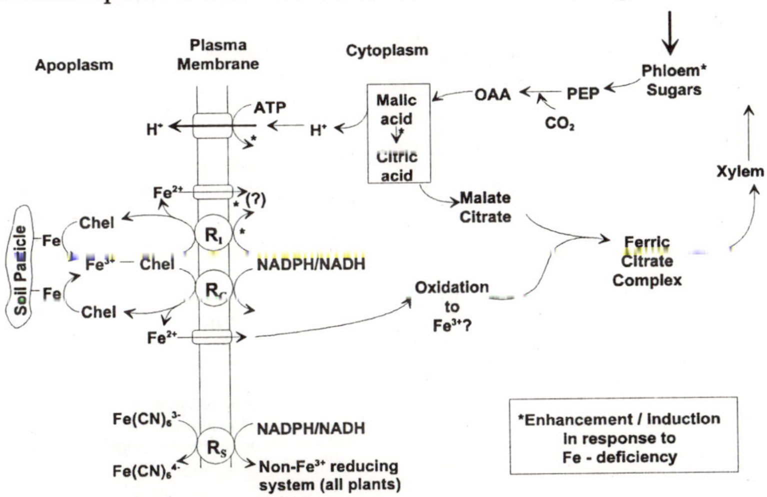

Figure 3.13. ***Ferric-chelate reduction-based model depicting the various physiological processes thought to be involved in the reduction of Fe(III) at the root-cell plasma membrane, and the subsequent absorption of Fe(II) ions into the root-cell of dicot and non-graminaceous monocot plants. Central to this model is the inducible Fe(III) reductase (Ri) in the plasma membrane that is induced in response to Fe deficiency. A constitutive Fe(III) reductase (Rc) is hypothesized to function under Fe-sufficient conditions.***

In grasses, Fe uptake is enhanced by a different mechanism, one that relies on the plant production of phytosiderophores. Phytosiderophore is a term used to describe plant produced chelates or complexing material that can increase the availability of Fe^{+++}. An illustration of this mechanism is provided in Fig. 3.14.

Many organisms use Fe^{+++} as an electron acceptor such as some fungi and chemoorganotrophic or chemolithtropic bacteria. This bacterial reduction of ferric to ferrous is a major way iron is solubilized. Reduction takes place under anaerobic conditions (waterlogged). *Shewenella putrefaciens* is one organism capable of reducing iron. Oxidation occurs under aerobic conditions. At neutral pH, organisms such as *Gallionella ferruginea* or Leptothrix oxidize iron. Under acidic conditions, *Thiobacillus ferrooxidans* is the primary organism responsible for iron oxidation. This organism is typical in acid mine drainage areas.

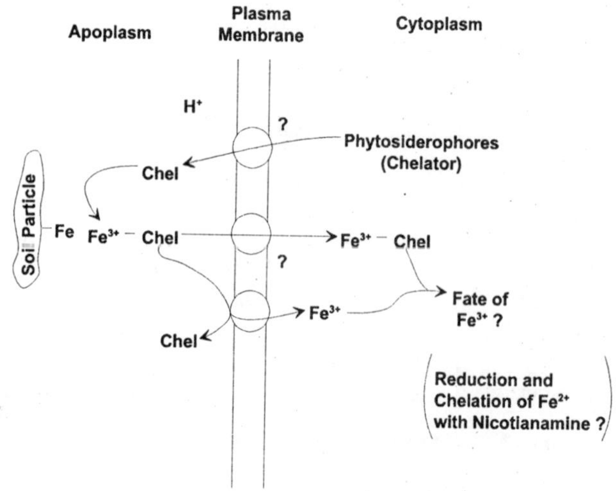

Figure 3.14. ***Phytosiderophore-based model for Fe absorption in grasses.***

A common iron mineral in nature is pyrite (FeS_2). Pyrite is often associated with bituminous coal and other ores. Bacterial oxidation of pyrite generates acid and is the cause of acid mine drainage.

$$FeS_2 + 31/2O_2 + H_2O \rightarrow Fe^{+2} + 2SO_4^{-2} + 2H^+$$

Fe^{++} hydrolyzes to form hydrolysis products common under reduced conditions. $FeOH^+$ predominates in solution at pH< 6.75, while $Fe(OH)_2^0$ prdeominates at pH >9.3. Magnetite (Fe_3O_4) is a stable mineral under reduced conditions.

Iron availability is increased greatly in acid soils, consequently the deficiency is seldom observed in any crops, where soil pH is usually less than 7.0. Iron deficiency cannot be corrected by soil application of iron containing fertilizers because the iron from the fertilizer is quickly converted to unavailable iron just like that already present in the soil. The exception to this general rule is the use of chelated iron. However, these fertilizer materials are cost prohibitive for field scale use. Foliar application of iron sulphate solutions is effective for supplying iron to deficient plants. Unfortunately, iron is not translocated in the plant and subsequent new leaves will again exhibit the interveinal chlorosis (yellow between the veins) so characteristic of iron deficiency. Repeated spraying will provide iron for normal growth but will often be cost prohibitive. The most effective long-term corrective measure for dealing with iron chlorosis is to increase soil organic matter since iron deficiency is usually limited to small areas of a field. Organic matter can be effectively increased by annual additions of animal manure or rotted hay. This results in additional chelating of iron and also has a tendency to acidify the soil. Broadleaf plants have what is called an "adaptive response mechanism" that allows them to make iron more available if they experience iron stress. The strength of this mechanism is a genetic trait and some varieties, do not possess this ability and will often become chlorotic if grown in neutral or alkaline soils.

Zinc

Zinc composition of soils defined by parent material. Magmatic rocks have 40 and 100 mg/kg Zn in granites and basalt, respectively. Sedimentary rock composition varies in the range 10 to 30 mg/kg in sandstones and dolomites, and 80-120 mg/kg in clays. Total content of Zn in soils ranges from 10 to 300 ppm (0.001-0.03%). In soils, 30-60% Zn can be found in iron oxides, 20-45% in the lattice of clay minerals, and 1-7% on clay exchange complex. Highest Zn concentration is in solonchaks - saline soils in Asia, lowest in light textured soils with low organic matter.

Soil Zn reactions and availability

Availability is closely linked to soil pH and organic matter content. Concentration of total Zn increases with depth, whereas extractable Zn content decreases. Concentration of Zn in the upper horizon also depends on organic matter content, which can hold up to 13% Zn. pH is the most important parameter of Zn solubility. General equation for soil Zn is -

$$pZn = 2pH - 5.8$$

The form of Zn predominant at

$pH < 7.7$ - Zn^{++}

$pH > 7.7$ - $ZnOH^{+}$

$pH < 7.7$ - $Zn(OH)_2$

Zn has only one oxidation state as an ion, Zn^{++}. Zn^{++} is immobile in both the plant and soil. Zinc is absorbed by plants as the divalent cation Zn^{++} and transported as citrate, malate and malonate complexes. Mobility in plants does not coincide with water flow.

Zn functions as an ion for coupling enzymes and substrate. The most common Zn containing enzyme is alcohol dehydrogenase. Depending on genotype, Zn concentration varies in the range 25-150 ppm (0.0025-0.015% of dry weight) of Zn sufficient plant. The critical deficiency concentration is from 15 to 30 ppm, higher if leaf P is above normal.

Interaction of Zn with other nutrients. Increase in available P content can considerably decrease availability of Zn in the soil due to the high antagonism between these two elements. However, some authors suggest that symptoms considered as a Zn deficiency are actually P toxicity. Presence of other nutrients such as iron, copper, manganese and calcium may also inhibit

Zn uptake by plants, probably due to the competition for the carrier sites on roots.

Deficiency of Zn found in acidic, sandy soils with high leaching, calcareous soils (pH>8.0) and exposed subsoil horizons (erosion). Application of high rates of NPK fertilizers can aggravate Zn deficiency. Deficiencies may occur whenever DTPA soil test values are less than 2.0 ppm. Deficiencies can be easily corrected by broadcast application of about 5 to 6 kg/ha of zinc sulphate. An application of this rate should remove the deficiency for 3 to 4 years. Foliar sprays are very effective in preventing and/or correcting the deficiency, a single application usually lasting the entire growing season.

Most plant species have high tolerance to excessive amounts of Zn. However, on acid and heavily sludged soils Zn toxicity can take place. Zn toxicity symptoms are Inhibited root elongation, photosynthesis in leaves, depresses RuBP carboxylase activity, chlorosis in young leaves due to induced deficiency of Fe^{2+} and/or Mg^{2+}. Zn^{2+} has ion radius similar to Fe^{2+} and Mg^{2+}, which creates unequal competition for these elements when zinc supply is high. The critical toxicity level in leaves is 100-300 mg per kg of dry weight.

Role of Zn in microbial growth. Indispensability of Zn in metabolism of living organisms, microflora also is highly dependent on concentrations of zinc present. Some heterotrophs can tolerate high concentration of Zn and behave as bioaccumulators of Zn, among them Zoogloea-producing bacteria, Ephiphytic bacteria, Nonsporing bacteria. Different genera of Green Algae respond differently to Zn contamination. *Microspora, Ulothrix, Hormidium,* and *Stigeoclonium* are resistant to high Zn concentrations, whereas genera such as *Oedogonium* and *Cladophora* are rather sensitive to the presence of Zn.

Manganese

Mn concentration in soils varies widely (20 to 3,000 ppm) depending upon parent material and the degree of soil weathering. Manganese is a cofactor in many plant reactions. It is essential for chloroplast production. The most widely known function is probably its involvement in the photosynthesis, wherein there is a $4e^-$ transfer that results in the splitting of water and release of O_2.

$$2H_2O + 4e^- \longrightarrow 4H^+ + O_2$$

Manganese (Mn) is adsorbed by plants in the ionic form as divalent cations Mn^{++}. Mn is easily oxidized from the Mn^{++} to Mn^{+++} and Mn^{++++}, the Mn^{++} is not strongly chelated, while the Mn^{++++} may be strongly complexed. Oxides of the highest valence state are quite insoluble, hence availability can be greatly affected by redox potential.

Soil Mn reactions and availability

Mn decreases 100-fold for each unit increase in pH. Concentration of Mn^{2+} in solution is increased under acid, low-redox conditions. High pH promotes the formation of less available organic complexes. Activity of soil microorganisms that oxidize soluble Mn to unavailable forms reaches a maximum near pH 7.0. High pH favours oxidation to Mn^{++++}, from which insoluble oxides are formed (MnO_2, Mn_2O_3, and Mn_3O_4). At pH < 6.0, reduction of Mn and formation of more available divalent form Mn^{++} takes place.

Poor aeration increases Mn availability and soil waterlogging reduce O_2 and lower redox potential, which increases soluble Mn^{++}. Local accumulation of CO_2 around roots increases Mn availability. High organic matter (particularly if basic soil) forms unavailable chelated Mn^{++} compounds, particularly in peat and muck soils. Pronounced seasonal variations, with wet weather increasing Mn^{++} and warm, dry weather encouraging the formation of less available oxidized forms.

Mn deficiency and toxicity

Deficiencies are most common in highly weathered soils that have been recently limed. Some deficiencies are caused by soil organisms oxidizing Mn^{2+} to Mn^{3+} ; Mn- efficient and Mn-inefficient plants. Liming and burning can produce alkaline conditions causing deficiency. Dry soils allow rapid oxidation and deficiency of Mn may takes place. Mn uptake can be improved in some plants if they undergo Fe stress and are able to respond by producing a reducing agent, since it will reduce both iron and manganese to more soluble oxidation states.

High levels of Cu, Fe or Zn can reduce Mn uptake. High levels of Mn can reduce Fe concentrations and induce Fe deficiencies and vice versa; ratio of Fe to Mn should be between 1.5 to 2.5. Mn and Al toxicities frequently occur together on acid soils.

Copper

Copper is a component of enzymes involved with photosynthesis. Of the divalent cations, Cu^{++} is the most strongly complexed by organic matter, oxides of Fe, Al, Mn, phenolic carboxyl, and hydroxyl groups, and clay minerals. Cu can be isomorphically substituted for Fe or Mn and leach through the soil profile in humus-poor, acidic peat, or in very acidic mineral soils, such as those around Ni and Cu smelters. Concentration of natural Cu in soil is 34 to 55 ppm.

Cu is absorbed as Cu^{++} and is relatively immobile in the plant. Because it undergoes oxidation-reduction reactions relatively easily, Cu is involved in electron transfer and enzyme systems much like Fe, most notably the oxidase enzymes.

The critical deficiency concentration is about 1 to 3 ppm. Typical deficiency symptoms are chlorosis (white tip), necrosis, and die-back in the youngest leaves. Deficiencies are most common in high organic (peat) soils because the Cu in the soil is bound too tightly for plants to extract adequate amounts. As much as 98% of all the Cu^{++} in the soil solution may be present as organic complexes.

Molybdenum

Molybdenum functions in plants in the enzyme nitrate reductase, which is very important in nitrogen metabolism in legumes. Mo found in plants primarily as the oxyanion (oxidation state VI), but also as Mo (V) and Mo(IV). Mo is absorbed as oxyanion MoO_4^{2-}, since it is the dominant species above pH 4.5 (Fig. 3.15).

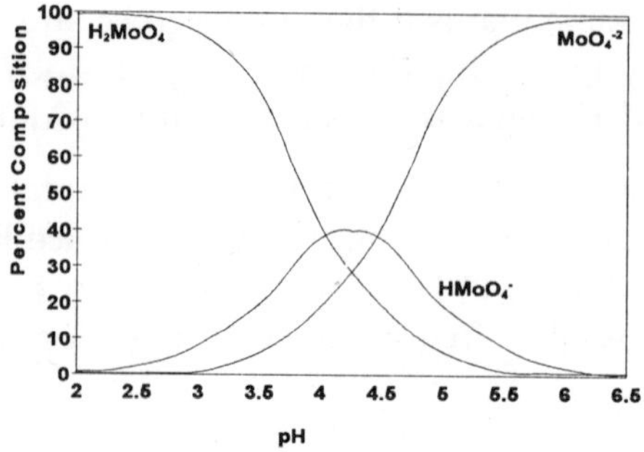

Figure 3.15. ***Relationship of molybdate ion species to pH.***

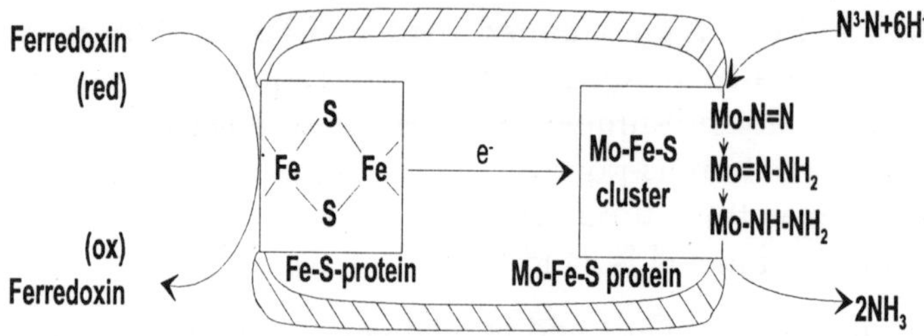

Figure 3.16. *Structural model of the nitrate reductase with its two subunits. Each subunit contains three prosthetic groups: FAD, heme-Fe, and Mo-protein.*

Mo functions in electron transfer in plants, primarily in nitrate reductase (Fig. 3.16) in non-legumes and nitrogenase (Fig. 3.17) in legumes. In each case N reduction is involved. Plants supplied with NH_4^+ have a much lower demand for Mo.

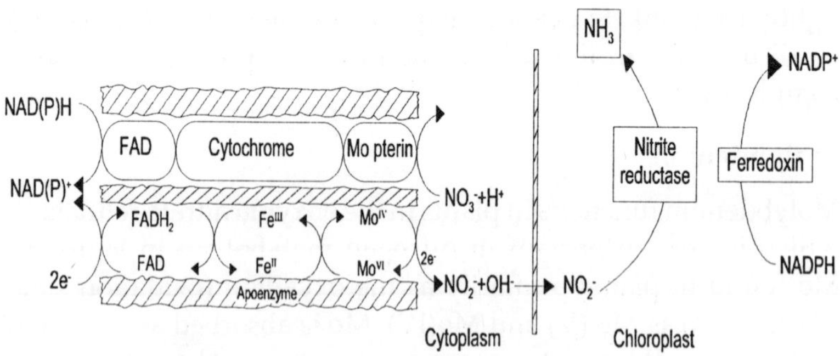

Figure 3.17. ***Model of the stepwise N_2 reduction by the Mo-containing nitrogenase.***

The normal concentration of Mo is quite low, ranging from about 1 to 10 ppm. The critical deficiency level ranges from 0.1 to 1 ppm in leaves, whereas critical toxicity concentrations range from 100 to 1,000 ppm. Deficiencies are uncommon, but are more likely in acid than alkaline soils apparently because MoO_4^{2-} is strongly adsorbed to iron oxide surfaces in acid soils, either as a result of chemical bonding or simple anion exchange associated with pH dependent charges in acid soils.

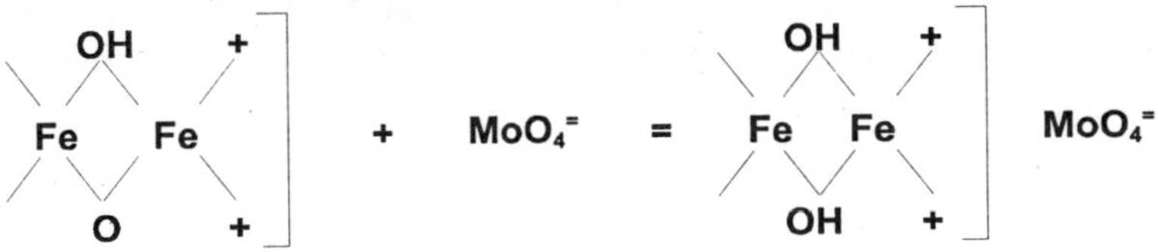

Bonding mechanisms

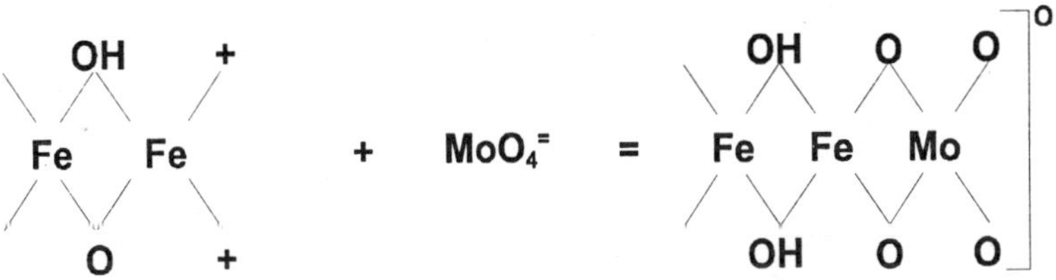

Exchange mechanisms

Availability is reduced in acid soils and often if molybdenum availability is marginal, it can be increased to adequate levels by simply liming the soil. Where large seeded legumes are grown like soybeans or peanuts, obtaining seed that was grown with a good supply of Mo will avoid the deficiency because normal levels of Mo in the seed will be enough to meet the plant needs.

P additions increase Mo uptake by replacements on the exchange complex and release to solution. Sulphur depressed Mo uptake by direct competition on root adsorption sites. Mo, with Mn, affects Fe uptake in tomatoes.

Mo is readily translocated and deficiency symptoms are normally found in the oldest leaves. General deficiency symptoms are varied between plants and range from yellowing, stunting, and cupping of older leaves followed by necrotic spots at the tips and margins. For legumes the symptoms are like that for N deficiency. In non-legumes the condition of "whip tail", where leaf blades are reduced and irregularly formed is common together with interveinal mottling, marginal chlorosis, and accumulation of NO_3.

Boron

Boron (B) is absorbed by plants as uncharged boric acid, $B(OH)_3$, the chemical form also present in soil solution. The main function of boron is in cell growth and formation. The action appears to be in binding sugars together in cis-diol ester linkages, resulting in B being strongly complexed in cell walls. An example of this is shown in the following generalized reactions.

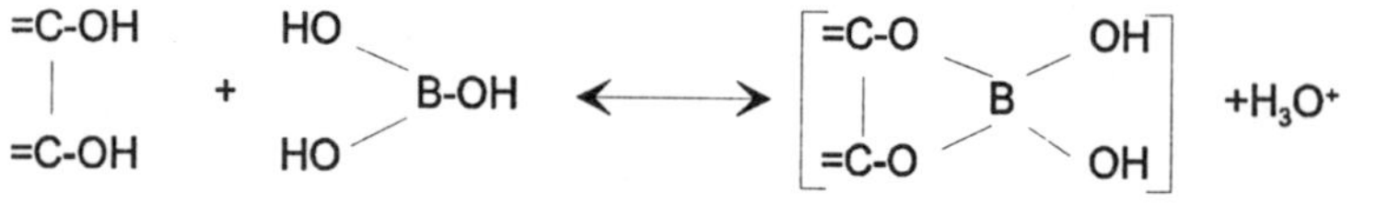

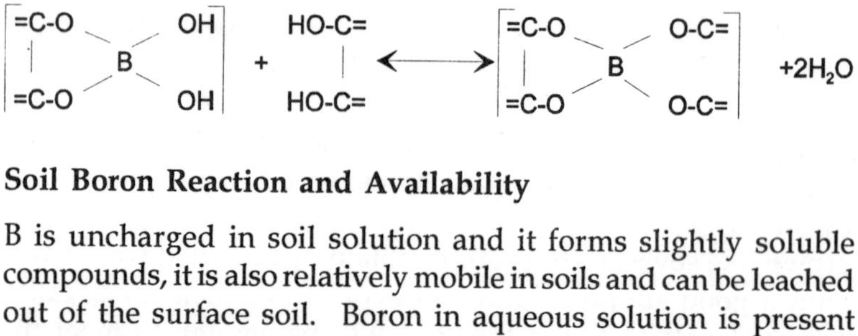

Soil Boron Reaction and Availability

B is uncharged in soil solution and it forms slightly soluble compounds, it is also relatively mobile in soils and can be leached out of the surface soil. Boron in aqueous solution is present mainly as undissociated boric acid, $B(OH)_3$. It dissociates according to the equation:

$$B(OH)_3 + HOH -- B(OH)^-_4 + H^+ \; K_{eq} = 10^{-9.2}$$

$$[B(OH)^-_4] [H^+]/[B(OH)_3] = 10^{-9.2}$$

and, rearranging we have

$$[B(OH)^-_4] / [B(OH)_3] = 10^{-9.2}/[H^+]$$

taking the log of both sides, results in

$$\log [B(OH)^-_4] / [B(OH)_3] = -9.2 - \log [H^+]$$

or

$$\log [B(OH)^-_4] / [B(OH)_3] = -9.2 + pH$$

and at pH 7.2,

$$\log [B(OH)^-_4] / [B(OH)_3] = -2,$$

so there is 100 times less $B(OH)^-_4$ than $[B(OH)_3]$ (the ratio is 0.01), verifying that $B(OH)_3$ is the predominate B species in the soil solution of agricultural soils. Hence, unlike all other nutrients plants obtain from the soil, B is apparently taken up as the uncharged $B(OH)_3$.

Boron is complexed by O.M. and can be a major source of B to plants. Mineralization of O.M. releases boron to soil solution. The mineral source of boron in soils is Tourmaline, which is a very insoluble borosilicate mineral. Boron availability decreases with increasing pH. Over liming acid soils can cause boron deficiency because of interaction with calcium.

Boron is generally less available on sandy soils in humid regions, because of more leaching. This is especially true in acid soils with low organic matter (O.M.). Boron availability increases with increasing O.M. Most alkaline and calcareous soils contain sufficient Boron because the primary boron minerals have not been highly weathered and, more important, B products of weathering (H_3BO_3) have not been leached out as in humid region soils.

Boron deficiency and toxicity

Boron deficient plants exhibit a wide range of deficiency symptoms, but the most common symptoms include necrosis of the young leaves and terminal buds. Structures such as fruit, fleshy roots and tubers may exhibit necrosis or abnormalities related to the breakdown of internal tissues. Critical deficiency concentration ranges from 5-10 ppm in monocotyledons to 50-70 ppm in dicotyledons, to as high as 100 ppm in latex producing plants such as dandelion.

Boron deficiency may be critical in peanut production because of the very sandy, porous soils peanuts are produced in. The deficiency manifests itself as a condition known as "hollow heart" whereby the centre of the nut is not completely filled and an inferior crop is harvested. Whenever B deficiencies are suspected, and if B fertilizer is applied, care should be exercised as toxicities can be created by simply doubling the recommended rate.

Boron can be toxic on some alkaline soils when soil test or extractable boron exceeds 5 ppm. Irrigation water that contains more than 1 ppm, boron can also produce toxicity. The critical toxicity concentration is not much higher than the deficiency concentration. In corn it is about 100 ppm and cucumbers 400 ppm.

Chlorine

Cl anion is very soluble in most soils. Most Cl in soil comes from salt trapped in parent material, marine aerosols, and volcanic emissions. Most often found in apatite, hornblende, and some feldspars. Cl accumulation takes place where internal drainage of soils is restricted and in shallow groundwater where Cl can move by capillary action into the root zone and be deposited at or near the soil surface. It is rapidly cycled through soil systems due to mobility (except in extremely acid soils). Exchangeable Cl can occur in acid, kaolinitic soils which have pH dependent positive charges. In humid climate zones, Cl is leached through the soil system and in Arid to Semi-arid zones, it is concentrated in the soil horizon.

Chlorine is absorbed as the simple Cl^- anion. Cl functions in plants mainly as a mobile anion in processes related to osmotic pressure regulation (stomatal openings) and thereby lowers the availability of water cation transport to plants.

Average concentration in plants ranges from 1 to 20 g/kg (0.1 to 2%), while the concentration required for optimum growth ranges from 150-300 mg/kg (0.015-0.03%). General crop requirement is about 1 unit Cl for 10,000 units of dry matter produced, or about 2-3 kg/ha.

Deficiency symptoms include reduced root growth, wilting and curling of leaves and leaflets, bronzing and chlorosis similar to that for Mn deficiency. Recent studies suggest the critical toxicity concentrations range from 3-5 g/kg in sensitive plant species and 20-40 g/kg in tolerant plant species.

Uptake of NO_3 and SO_4 can be reduced by the competitive effects of Cl. Lower protein concentrations in winter wheat are attributed to strong competitive relationships between Cl and NO_3 when Cl levels are high. Negative interaction between Cl and NO_3 has been attributed to competition for carrier sites at root surfaces.

Soil fertility research has occasionally shown small grain response to fertilizers containing chlorine. Often the response has been the result of disease suppression rather than correction of an actual nutrient deficiency in the plant, and usually it has been in areas that do not commonly apply potassium fertilizers containing chloride (such as muriate of potash or potassium chloride).

The Mobility Concept

The nutrient mobility concept as it relates to soil fertility was first proposed in 1954 by Roger H. Bray at the University of Illinois. Much research since then has supported his mobility concept and it is now considered basic to the understanding of soil fertility. Bray simplified all the soil chemistry surrounding the essential nutrients to the fact that some are quite mobile in soils and others are relatively immobile.

Mobile Nutrients

Plants are able to extract mobile nutrients from a large volume of soil, even soil beyond the furthest extension of their roots because as the plants extract water from around their roots, water from further away moves toward the root and carries the mobile nutrient with it. Figure 3.18 illustrates this point. Plants obtain mobile nutrients from a "root system sorption zone" and are capable of using nearly all of the mobile nutrient (or mobile form of the nutrient) if the supply is limited.

According to Bray, the mobile nutrients are: Nitrogen, Sulphur, Boron, and Chlorine.

In a field situation, where there is more than one plant, root system sorption zones overlap if plants are close enough together as illustrated in Figure 3.19. As a result, there is a volume of soil between plants where the nutrient is in demand by both plants. As plants are placed closer and closer together (e.g. increasing plant population to increase potential yield) the competition for nutrients increases. Unless the competition among plants in a field for a mobile nutrient is satisfied by supplying more of the nutrient, the growth and yield of plants will be restricted.

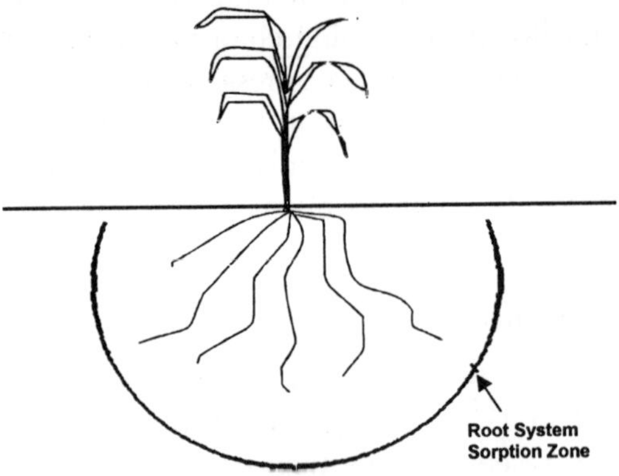

Figure 3.18. ***The large volume of soil from which plants extract mobile nutrients (root system sorption zone).***

Supply of mobile nutrients like nitrogen must be provided in direct proportion to the number of plants, or potential yield of the crop. This "supply" can be easily determined by calculating the amount of nutrient that will be taken up by the crop. In order to do this, we only need to know the average concentration of the nutrient in the crop and what the crop yield will be. Average nutrient concentrations are commonly known, however, yields vary from field to field and year to year. For this reason, it is critical to have in mind a "yield goal" or expected yield in order to determine fertilizer needs for mobile nutrients like nitrogen. Because mobile nutrients are almost completely extracted from the root system zone, soil test values like nitrate nitrogen will change drastically from one year to the next in relation to how much nitrogen was available and the crop yield.

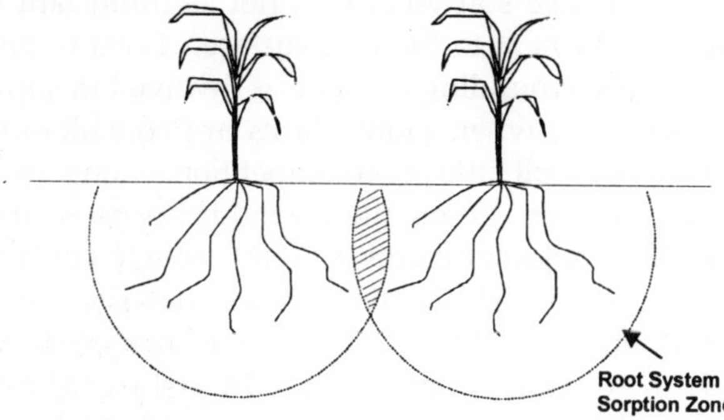

Figure 3.19. ***Competition among plants due to increasing yield goal.***

Immobile Nutrients

Nutrients that are immobile in the soil are: phosphorus, potassium, calcium, magnesium, iron, zinc, manganese, copper and molybdenum. These nutrients are not transported to plant roots as soil water moves to and is absorbed by the root. These nutrients are absorbed from the soil and soil water that is right next to the root surface. Because of this, there is only a small volume of soil next to the root surface that is involved in supplying immobile nutrients to plants.

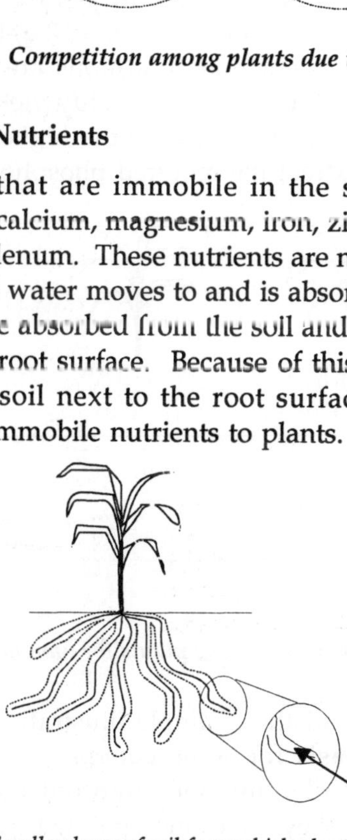

Figure 3.20. *Small volume of soil from which plants extract immobile nutrients (root surface sorption zone).*

Figure 3.20 identifies this soil volume as the "root surface sorption zone". This figure illustrates that only a small fraction of the soil in the total rooting zone is actually involved in supplying immobile nutrients. The total amount of immobile nutrient in the whole soil volume is not as important as the concentration right next to the root surface. Because only the thin layer of soil surrounding the roots is involved in supplying immobile nutrients, when more plants are considered as in Figure 3.21, there is still little or no competition among the plants for immobile nutrients. Competition would only occur at points where roots from adjacent plants actually came in contact with one another. This illustration indicates that the supply of immobile nutrients like phosphorus does not have to be adjusted (increased) in relation to an increase in yield goal or yield potential. If soil availability is adequate for a 12 q/ha of paddy crop yield, then in the event that conditions are favourable (better moisture supply) for 24 q/ha of paddy yield, the more extensive root system that develops for the higher yield will explore new soil and extract the required phosphorus.

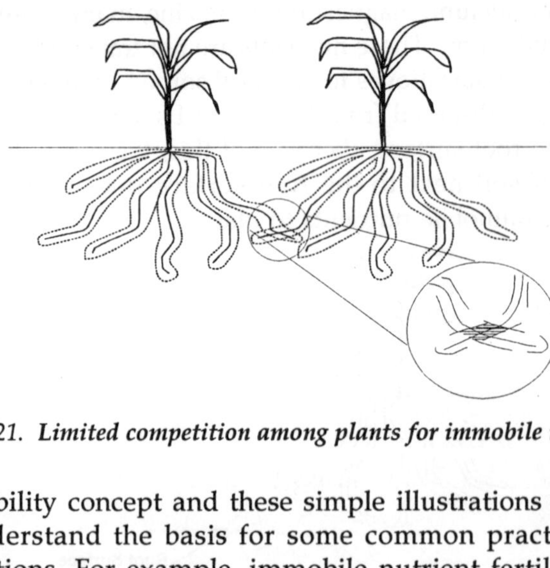

Figure 3.21. ***Limited competition among plants for immobile nutrients.***

The mobility concept and these simple illustrations can help one understand the basis for some common practices and observations. For example, immobile nutrient fertilizers are usually more effective if they can be incorporated, but especially should be placed where roots have a high probability of coming in contact with the fertilizer. This is why band applying phosphate fertilizers is generally more effective than the same rate broadcast and incorporated. Mobile nutrients like nitrogen can be broadcast during the growing season (topdressing) because they are easily moved to the roots with rain or irrigation. The phosphorus soil test does not change much from year to year regardless of the previous year's yield or fertilizer rate because much of the soil is not in contact with the roots or fertilizer and its available phosphorus status is therefore unchanged. Continued broadcast application of high rates of phosphorus may cause a build up and an increase in the soil test phosphorus because only a fraction (15 to 20%) of the fertilizer comes in contact with the roots (fertilizes the crop) and the rest reacts only with the soil (fertilizes the soil).

It is sometimes useful to compare mobile and immobile nutrients and their management to fuel and oil for a tractor or pickup. Fuel is required in relation to the amount of work expected from the tractor in much the same way nitrogen is required in relation to the amount of yield expected from the crop. Oil is required more in relation to the level in the crankcase identified by the dipstick than by what or how much work is expected from the tractor (oil burners excepted). Similarly, phosphorus and potassium requirements are determined from the soil test and the amount of fertilizer recommended does not depend on the yield goal. Like the dipstick that is calibrated with a mark showing "full" and "1-quart" low, the soil test for phosphorus (and any immobile nutrient) must be calibrated by field research. Just as the dipstick is uniquely calibrated for each kind of tractor, soil test calibrations vary slightly for different crops and soils and may be somewhat unique for states and regions.

4

Soil-Water-Plant Relationship

Plant growth depends on the use of two important natural resources, soil and water. Soil provides the mechanical and nutrient support necessary for plant growth. Water is essential for plant life processes. Therefore, soil, water and plant relationship is essential to maintain the soil water between these extremes, allowing nutrient movement, aeration, and supplying water in excess of evaporation and transpiration (evapotranspiration). Measuring plant available water and adjusting water levels with irrigation is another way to modify the environment to maximize food and fibre production.

HOW SOIL HOLDS WATER

Soil holds water in two ways, as a thin film on individual soil particles, and as water stored in the pores of the soil. Water stored as a thin film on individual soil particles is said to be in *adsorption*. Adsorption involves complex chemical and physical reactions but in simple terms, a thin film of water adheres to the outside layers of soil particle molecules. Water stored in the pores of the soil is said to be in *capillary storage*.

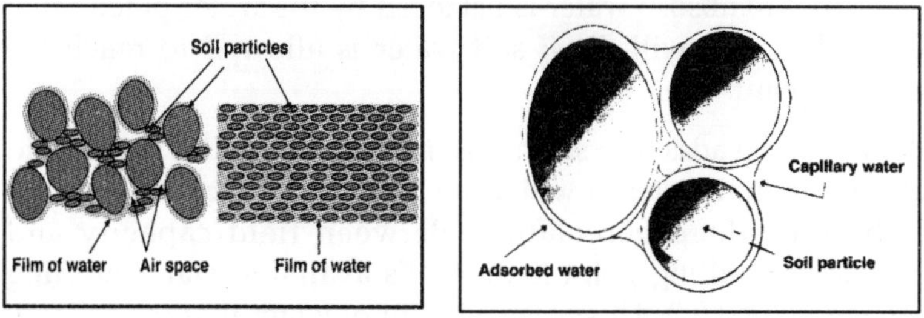

Figure 4.1. *Pore spaces and soil particles.*

An example of this phenomenon would be to place one end of a glass capillary tube in a pan of water. Water in the tube will rise to a certain height, which depends on the diameter of the capillary tube. This phenomenon can act in any direction and is the key to water being stored in soil pores.

SOIL AND WATER INTERACTIONS

It is important to understand the interactions between the soil and water, which include soil water content, how the soil holds the water, and soil water tension.

Soil Water

Soil "holds" water against the pull of gravity, retaining it for crop use. There are limits to this ability. The upper limit is called **field capacity**. The lower is called **permanent wilting point**.

At saturation, which usually occurs immediately after a heavy rainfall or irrigation, all pore spaces in the soil are filled with water. When the soil is at or near saturation, some of the water is free to drain or percolate due to the force of gravity. This excess water is referred to as **gravitational water**. Since this percolation takes time, some of this extra water can be used by plants or lost to evaporation.

Field capacity is defined as the amount of water remaining in the soil after percolation has occurred. This is not a very definite soil water point; therefore, field capacity often is defined as approximately one-third atmosphere tension. Wilting point is defined as the soil water content at which the potential of the plant root to absorb water is balanced by the water potential of the soil. Crops will die if soil water is allowed to reach the wilting point.

Water can be added to soil above field capacity. It will not hold this additional water. It will drain below the effective root zone to become deep percolation. Between field capacity and permanent wilting point is the soil's **available water holding capacity** (AWHC). This is the amount of water that the soil will hold that is available for the crop to use.

As the crop pulls water from the soil, the soil holds onto the remaining water harder and harder, putting more and more stress on the crop. Crop wilt during the hottest parts of the day. Sooner or later, if the stress gets big enough, the crop will permanently wilt as soil moisture is at permanent wilting point.

Soil water can be measured in two ways:

a) Volumetrically, the actual amount of water in the soil.

b) Tension, a measure of the water holding forces in the soil.

The crop doesn't care how much water is physically in the soil, only how hard it is to get out. Thus, although a volumetric measurement will tell us how much water is in the soil and, therefore, how much to irrigate, the tension measurement is more important in terms of preventing crop stress. The volumetric standard of measurement is inches of water held per foot of soil. Available water holding capacities vary from 1 to 2.5 inches/foot. The tension standard of measurement is pressure, usually centibars.

INFILTRATION RATE AND APPLICATION RATE

The rate at which water soaks into the soil is called the "**infiltration rate**". The infiltration rate, measured in inches or cm/hour. Infiltration rates will decrease during an irrigation.

The rate at which we apply water through irrigation, or the rate that water falls during a rain storm is called the "**application rate**". The application rate, also measured in inches or cm/hour. Surface runoff occurs when the application rate of the irrigation/ rainfall is greater than the infiltration rate of the soil.

Water applied to the soil by precipitation (natural or manmade, such as sprinkling systems) infiltrates; and some of the water may be stored temporarily on the soil surface if the soil is unable to absorb it. Thus, if the rate of application exceeds the infiltration rate, water collects on the surface, and either ponding takes place or the water runs off. The infiltration rate governs the amount of water entering the soil and the amount that can be stored in the soil profile to be available for crops. In addition, the infiltration rate governs the amount of potential runoff and its associated soil erosion threat. As an example, surface sealing or crusting can reduce infiltration, increase erosion, and limit the available water for plants.

Percolation

The infiltration rate is limited by the ability of the soil to transmit water away from the soil surface through the soil profile when the surface is ponded. This movement of water through the soil profile is known as percolation. Percolation rate is governed by the permeability of the soil or its hydraulic conductivity. Both terms describe the ease with which soil transmits water.

Because water percolates chiefly through large pores in a soil, percolation depends on the relative number and continuity of these pores. Soil with high porosity and coarse open texture has high hydraulic conductivity. For two soils of the same total porosity, the soil with small pores has lower conductivity than the soil with large pores because resistance to flow is greater in small pores. Soil with pores of many sizes conducts water faster if large pores form continuous paths through the profile. In fine-textured soils, conductivity depends almost entirely on the pore space between structural units. In some soils, particles are cemented together to form nearly impermeable layers commonly called hardpans. In other soils, very finely divided or colloidal material expands on absorbing water to form an impervious gelatinous mass that restricts water movement.

Quality of water transmitted, particularly its salinity and alkalinity, may have a marked effect on hydraulic conductivity. Change in the viscosity of water has an effect. Chemical change in water may affect hydraulic conductivity greatly without changing viscosity. The addition of even small amounts of sodium chloride to the soil water, insufficient to make any noticeable difference in viscosity, may affect soil structure so much that hydraulic conductivity is greatly reduced.

There are several methods available for measurement of soil moisture, both for volumetric (the neutron probe, gravimetric, "feel") and tension (tensiometers, gypsum blocks, leaf pressure chambers). They all have their strengths and weaknesses. Very high or out-of-balance salts will modify many of the measurements and results of different measurements (high or low).

SOIL WATER POTENTIAL

Soil water is frequently described in terms of content in units of gravimetric percent, percent on a volume basis, or equivalent water depth per unit of soil depth. A descriptive property is needed, however, to explain why soils treated in similar ways have different water contents; why plants respond differently on contrasting soils even though they have the same water content; and why if a sandy soil and clay soil have the same water content and are placed in intimate contact with one another, water will move from the sandy soil to the finer textured soil. Soil water potential is the property used to describe such a phenomenon.

Total Water Potential

It is the amount of work required per unit quantity of pure water to transport, reversibly and isothermally, a small quantity of water from a pool of pure water at a specified elevation at atmospheric pressure to the soil water at the point under consideration. Differences in potential energy of water from one point in a soil system to another give rise to the tendency of water to flow within the soil. In the soil, water moves continuously in the direction of decreasing potential energy.

The concept of soil-water potential is of great fundamental importance. Water in the soil differs from place to place and from time to time, not in form, but in potential energy. For very practical reasons, however, it is convenient to retain the concepts of "field capacity" and "permanent wilting point," while recognizing the qualitative aspects of such nomenclature. Total water potential consists of several components. It is the sum of matric, solute, gravitational, and pressure potential:

Total Water Potential= Matric + Solute + Gravitational + Pressure

Units of the potential depend on how a unit quantity of water is specified. Because weight is one of the most convenient ways of specifying the unit of water.

Gravitational Potential

This component of total potential is independent of soil properties and depends only on the vertical distance between the reference and the point in question. For points above the reference, gravitational potential is positive; points below the reference are negative.

In Figure 4.2, two points in a soil are located at a specific distance from a reference point Z. Gravitational potential of point A is 6 inches and of point B is 4 inches, thus the difference in gravitational potential between the two points is 6 inches - (-4 inches) = 10 inches.

Matric Potential

When the unit quantity of water is expressed as weight, then matric potential is the vertical distance between the measured point of the soil (ceramic cup of Fig. 4.2) and the water surface of a water-filled manometer. Matric potential is a dynamic soil property and will be at a theoretical zero level for a saturated soil. The matric potential of a soil system results from capillary and adsorptive forces due to the soil matrix. These forces attract and bind water in the soil and lower its potential energy below that of bulk water. Capillarity results from the surface tension of water and its contact angle with the solid soil particles. This potential was formerly called *capillary potential* or capillary water. In Figure 4.2, the unglazed ceramic cup that is embedded in soil is connected to a water manometer to form a tensiometer. The weight matric potential of the soil water at the cup is the vertical distance from the centre of the cup to the water level in the manometer which is 6 inches.

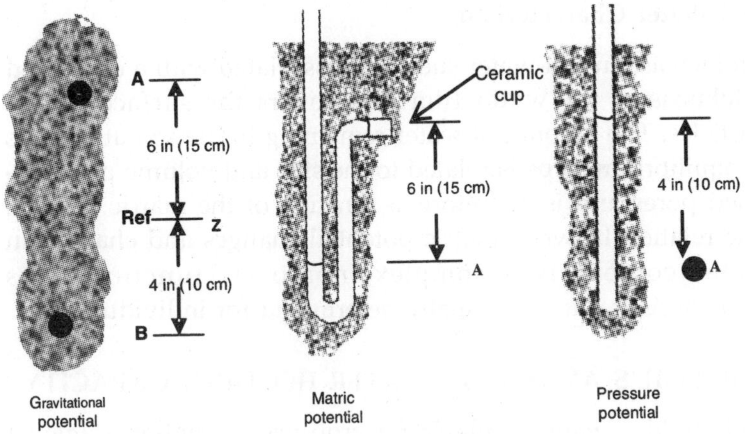

Figure 4.2. Gravitational, Matric and Pressure Potential.

Pressure Potential

The pressure potential applies mostly to saturated soils. Where water quantity is expressed as a weight, pressure potential is the vertical distance between the water surface and a specified point. In the field, this component is zero above and at the level of water in the piezometer. Below the water level it is always positive. In figure 4.2, a piezometer tube (tube open at both ends) is installed in the soil to a depth below the water table. Pressure potential at point A is the distance between the point and the water level which is 4 inches.

Solute Potential

Solute or osmotic potential arises because of soluble materials (generally salts) in the soil solution and the presence of a semi-permeable membrane. Two recognized membranes in soil-water systems are the cell wall of plant roots and air-water interfaces. The solute potential can be approximated from the relation:

$$\text{Solute potential} = RTC$$

where, R is the universal gas constant (82 bars cm^3/mol "k), T is absolute temperature ("k), and C is solute concentration (mol/cm^3).

Soil Water Characteristic

An increase in soil-water suction is associated with a decreased thickness of the water film that covers the surface of soil particles. The amount of water remaining in the soil at a series of equilibrium steps is related to the size and volume of water-filled pores and is, therefore, a function of the matric suction. The relation between matric potential changes and changes in water content is a complex, nonlinear function. This characteristic curve is usually determined for individual soils.

THE SOIL'S AVAILABLE WATER HOLDING CAPACITY

Soil "holds" water available for crop use, retaining it against the pull of gravity. This is one of the most important physical facts for agriculture. If the soil did not hold water, water free to flow downward with the pull of gravity as in a river or canal. The soil's ability to hold water depends on both the soil texture and structure. The finer the soil texture (higher percentage of silt and clay), the more water soil can hold.

Gravity is always working to pull water downwards below the plant's root zone. To counteract the pull of gravity, soil is able to generate its own forces, commonly called "matric forces" ("matric" because of the soil "matrix" structure that forms the basis for the forces).

An important fact about the soil's water-holding forces is that as the level of soil moisture goes down, the soil generates more force. This is the reason that some water will move up into the root zone from a shallow ground water table. As the plant extracts water in the root zone, the soil pulls water up from the area with more water to the area with less. The rate at which the water-holding forces go up with decreasing soil moisture is different for different soils. In a coarse soil, they will go up slowly. This means that plants can extract a great amount of water from coarse soils before they stress. In contrast, these forces rise quickly in finer soils.

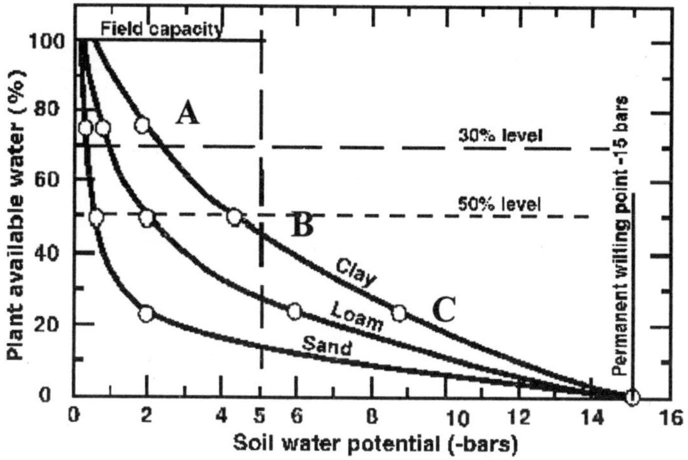

Figure 4.3. *Soil moisture level vs. Soil moisture tension.*

Graphically, the relationship can be described by the Figure 4.3. Looking at the lowest line for a coarse soil, it can be seen that at A, the soil moisture level is very high and the water-holding forces are low. This means that the plant can extract water easily from the soil. At B, the soil moisture level is lower but the water-holding forces haven't gone up that much. The plant can still extract water easily. However at C, the soil moisture level is very low and the water-holding forces have increased greatly. The plant cannot extract water easily and will be stressed.

Available Water and the Effective Root Zone

The water held by the soil between field capacity and permanent wilting point is termed the "available water holding capacity" of the soil. It is water that is "available" for the plant to use. Water added to the soil in excess of field capacity will drain down, below the active root system. Water held by the soil that is below the permanent wilting point is of no use, the plant will die.

Table 4.1. Normal crop rooting depths (cm).

Crop	Root depth	Crop	Root depth
Alfalfa	150-300	Grain	90-120
Asparagus	180-300	Grass	60-120
Beans	90-120	Lettuce	30-45
Beets (sugar)	120-180	Melons	180
Broccoli	60	Onions	30
Cabbage	60	Peas	90-120
Carrots	60-90	Peppers	60-90
Cauliflower	60	Potatoes	90-120
Celery	90	Spinach	60
Citrus	120-180	Squash	90
Corn (sweet)	90	Strawberry	90-120
Corn (field)	120-150	Tomatoes	180-300
Cotton	180-240	Walnuts	360
Deciduous tree	180-240		
Grapes	120-180		

The plant's active root system is called "effective root zone". The effective root zone is that depth of soil where soil moisture is to be maintained. The effective root zone may or may not be the actual depth of all active roots. It may be shallower because of concerns for crop quality or development (as with many vegetable crops). For a pre-irrigation though, there is need to consider the maximum potential root zone as the effective root zone for that irrigation. Table 4.1 shows some estimates for effective root zones healthy crops grown on deep well-drained soils. Note the wide range in some of the root zones in the table above. This is an example of irrigation as both an art and a science. There are scientific ideas like "effective root zones" and "field capacity" that provide a way to think about water management. But it is up to the grower, to apply these ideas, to pick the effective root zone, to estimate the field capacity and permanent wilting points.

MOVEMENT OF WATER INTO AND THROUGH THE SOIL

During an irrigation or rain, water soaks into the soil at a rate dependent on the soil type/structure, the soil/water chemistry, and the current soil moisture. Usually, the soil's infiltration rate decreases with time during an irrigation or rain. This is graphically illustrated in Figure 4.4.

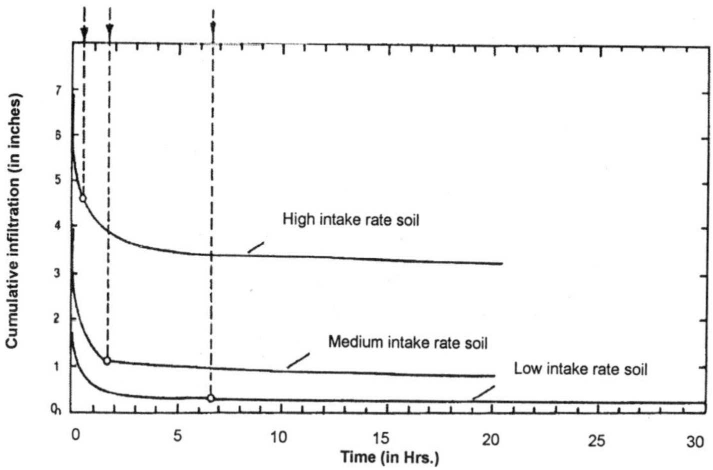

Figure 4.4. *Soil infiltration curves, Rate (in/hr) vs. Time.*

Figure 4.4 is a graph of the infiltration rate at the head of a furrow during irrigation. At the left, the irrigation has just started (elapsed time = 0) and the infiltration rate is high. The water is taking in very quickly. But 20 hours into the irrigation, the infiltration rate has dropped dramatically.

Water movement into and through soil is very much influenced by the soil/water chemistry and soil structure. A "sodic" soil will usually have an imbalance in salts that reduces the permeability of the soil.

Water moves through the soil due to a combination of gravitational and matric forces. Water always tends to move down. However, as described previously, soil water-holding forces are higher in areas with low soil moisture. Thus, water will also tend to move from soil with high soil moisture to areas of less. This is why water can move up into the root zone from a shallow ground water table. As the plant's root system extracts water in the immediate vicinity of the root, water from the wetter surrounding soil will move towards the drier area around the root. As the soil moisture goes down, less and less water will move towards the root and that water surrounding the root is held tighter. Thus, the root can extract less and less water and stress is put on the plant as the soil dries.

Runoff and the Irrigation Application Rate versus the Soil's Infiltration Rate

Infiltration rates and application rates are both measured in terms of inches of water applied per hour. For example, a standard field sprinkler system using 7/64" nozzles and running at 50 psi applies water at about 0.2 inches of water per hour. That is, for every hour the system runs, an equivalent depth of about 0.2 inches of water is sprayed onto the ground. If the soil's infiltration rate is greater than 0.2 inches/hour, then all water applied by the sprinkler system will soak in. If it is less, then water standing on the surface or running off.

Figure 4.5 is a graph of infiltration rate versus time during an irrigation. The infiltration rate at the start of the irrigation (the left side) is very high, while at near the end of the irrigation (the right side) it has dropped greatly. The straight horizontal line represents the application rate of a sprinkler system that could have been used during the irrigation, high and low rate. It is straight because the application rate of the sprinkler system doesn't change. It pumps out the same amount of water all through the irrigation.

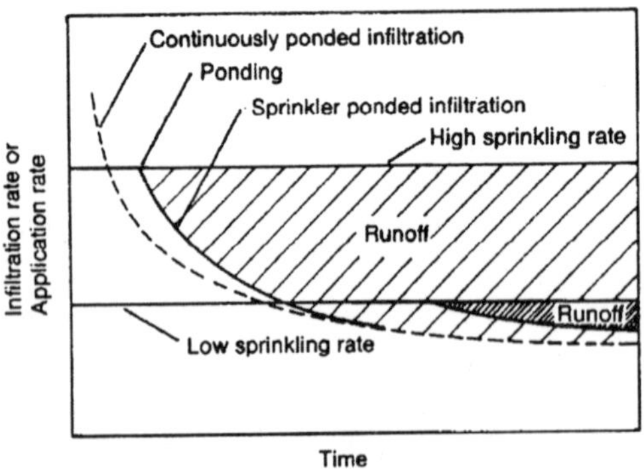

Figure 4.5. Sprinkler infiltration rates.

The infiltration rate is higher than the application rate at the start of the irrigation. Thus, all water applied by the sprinklers soaks into the soil. However, by the end of the irrigation, the infiltration rate has dropped below the application rate. Now the sprinkler system is applying more water than the soil can soak in. In this case, standing water or runoff occurs.

Water Movement in Soil under Saturated and Unsaturated Condition

Under saturated conditions, the rate of water movement in a soil system is governed by the characteristics of the pore space; therefore, the actual geometry and flow pattern of a soil is extremely complex. An equation known as *Darcy's law* is used to express the flux density (volume of water flowing through a unit cross-sectional area per unit of time).

The equation is:

$$q = K\,(dH)/L$$

where q is the flux density, (dH)/L is the hydraulic, and K is the proportionality factor force.

This law indicates that the flow of water through the soil is gradient (head drop or change in head per unit distance in the direction of, and at a rate proportional to, the driving direction of flow), and K is the proportionality factor force acting on the system generally designated as the hydraulic conductivity.

Many processes involving water movement reflects the complexities of individual soil systems. Crop root zone occur under unsaturated conditions. In comparison to saturated flow, water movement under unsaturated conditions is considerably more complex and difficult to describe quantitatively; however, many excellent quantitative reviews on this subject are available for definitive information.

Under unsaturated conditions, water in soils is subject to sub-atmospheric pressure that is equivalent to a negative pressure potential. A gradient of this potential constitutes a driving force. Matric suction, is due to the affinity of water to the soil-particle surfaces and capillary pores. Water is drawn from a region where hydration films are thicker to where they are thinner and from a zone where capillary menisci are less curved to where they are more highly curved. In other words, water flows from a region of higher to a region of lower matric potential (low to high suctions). For saturated flow, the hydraulic conductivity (K) of a given soil will remain constant. Under unsaturated conditions, K changes drastically with water content. Large, more highly conductive pores are drained first; therefore, K decreases dramatically as a given soil dries. Because different soils have varying pore space characteristics, K values also are markedly different; contrasting textural soil classes influence unsaturated hydraulic conductivity in a very pronounced way. Generally, a saturated sandy soil will conduct water more rapidly than a saturated clay soil. As the soils drain, the very opposite conditions prevail; the small pores of the clay soil will retain and conduct water even at appreciably low potentials. Under unsaturated field conditions, flow is much more pronounced with fine-textured soils than with sandy soils.

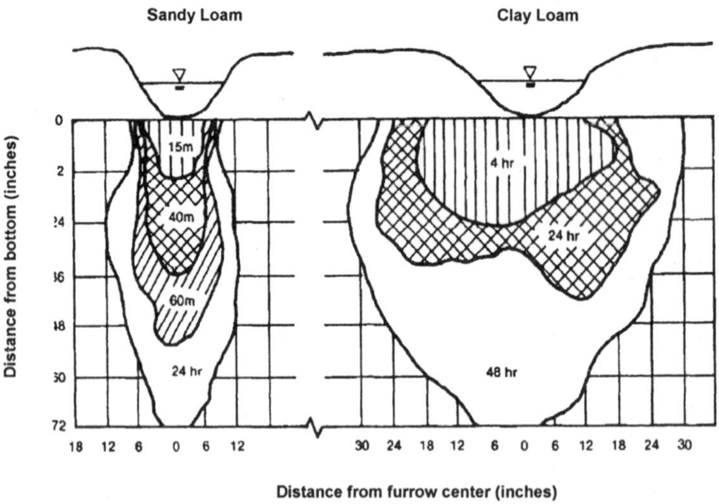

Figure 4.6. *Water penetration and movement in sandy and clay loam.*

From this discussion, it is evident that water moves in soils in any direction in relation to potential energy gradients. The rate and magnitude of movement is determined by the many and complex relations that exist in soil systems. A practical implication of textural class is illustrated in Figure 4.6. To achieve complete wetting, furrows have to be closer together on sandy soils.

PLANT AVAILABLE WATER

Traditionally, plant available water has been considered to be the amount of water held by the soil between field capacity (FC) and permanent wilting point (PWP). These two points provide only qualitative information on soil water retention properties; nevertheless, their usage continues and useful planning information can be obtained from these concepts. It is important, however, to understand the limitations that are imposed.

By definition, FC is the amount of water a well-drained soil holds after "free" water has drained off. For coarse textured soils drain off occurs soon after irrigation because of their relatively large pores. In fine-textured soils, drainage takes much longer because of their small pore size. Soil properties that affect field capacity materially are texture and strata within the profile that restrict water movement. Fine textured soils hold more water than coarse-textured soils. Field capacity for sandy soils is defined as -1/10 bar, for silty soils, -1/5 bar; and for clayey soils, -1/3 bar. Restricted flow in stratified soils slows redistribution, but may increase the amount of water used by the plant. The effect of contrasting soil texture on the soil-water potential is illustrated in Figure 4.7 by the broken line near the left vertical axis. Field capacity can be determined in the field after a soil has been thoroughly wetted by irrigation or rain or estimated in the laboratory from water-characteristic relations.

The permanent wilting point (PWP) is the soil-water content at which plants can no longer obtain enough water to meet minimal transpiration requirements; at which time, they wilt and if watered will not recover. Plants will wilt if they are not able to

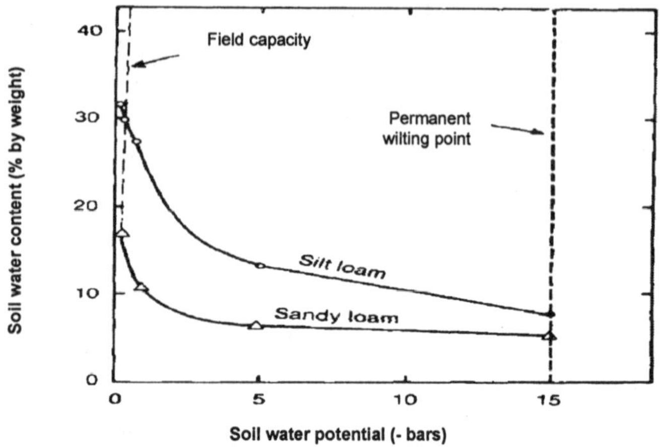

Figure 4.7. Relation between water content and water tension of a soil.

take up soil water fast enough to meet the climatic ET demand. Plants continue to absorb water when wilted, but not at a sufficient rate to regain turgor. The water potential commonly used for PWP is -15 bars, which was first established with sunflowers over a wide range of soils. This parameter is shown as the right vertical line of Figure 4.8. Some plants can extract soil water to potentials below -15 bars before they wilt and some will wilt above -15 bars.

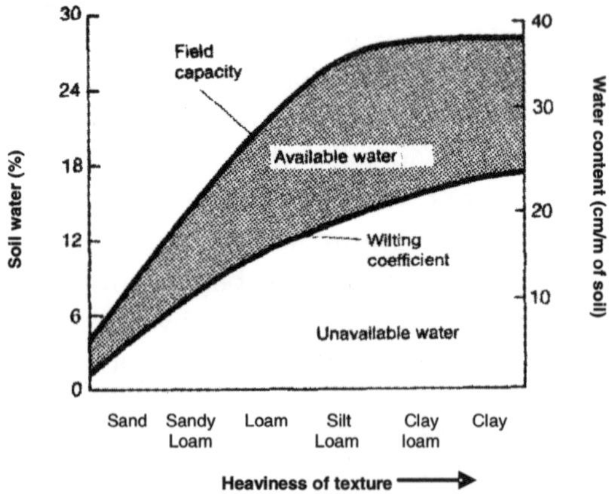

Figure 4.8. Relationship between soil water characteristics and texture.

Soil water considered to be available for plant growth lies at a potential energy level between FC and PWP. It should be pointed out, however, that these determined values represent only the matric potential of the soil water system. The presence of salts may contribute a substantial osmotic component to the total soil water potential. It is the total potential that determines soil water availability to plants. Nevertheless, it is helpful to examine some effects of soil texture on the water held between FC and PWP. For soils low in soluble salts, the finer the texture the greater the available water capacity (AWC).

Figure 4.8 shows the variation in FC and PWP water content by texture. Soil water content in percent by dry weight of soil is shown on the left margin and soil water content of soil is shown on the right margin for various soil bulk densities. The figure may be used as a general guide for estimating the AWC of soils based on texture until local curves can be developed. It applies generally to uniform soil profiles with low salt content.

Generally, well-drained sandy soils have a low available water capacity. Silty soils have a good available water holding capacity, as do clay loams and clays. Table 4.2 provides a general guide of available water ranges for given soil textural classes. While FC is considered to be the upper limit of available water, it should be pointed out that this is not strictly true. Water moving downward in the soil following an irrigation or rainfall can be effectively used by growing plants. Because this is a transitory stage, this water is generally not considered in calculations to determine the available water retaining capability of a soil but may affect irrigation scheduling.

Table 4.2. Ranges in available water content by soil textural classes.

Textural class	cm of water per cm of soil depth
Very coarse sand	0.03 - 0.06
Coarse sand-loamy sand	0.06 - 0.10
Sandy loam - fine sandy loam	0.10 - 0.14
Very fine sandy loam - silt loam	0.12 - 0.19
Sandy clay loam - clay loam	0.14 - 0.21
Sandy clay - clay	0.13 - 0.21
Peat and muck	0.17 - 0.25

Water held between FC and PWP is frequently considered to represent 100 percent of the available water supply. The water release characteristic curve in Figure 4.9 to illustrate this concept.

From Figure 4.9, it is clear that a given level of allowable depletion, for example, all growing plants require water daily. In irrigated regions, the depth of water to apply in each irrigation and the interval between irrigations are both influenced by storage capacity of the soil; therefore, the capacity of soils to store available water for use by growing crops is of special importance and interest. Irrigated soils that have large water-storage capacity may produce profitable crops in places where there is a shortage of irrigation water.

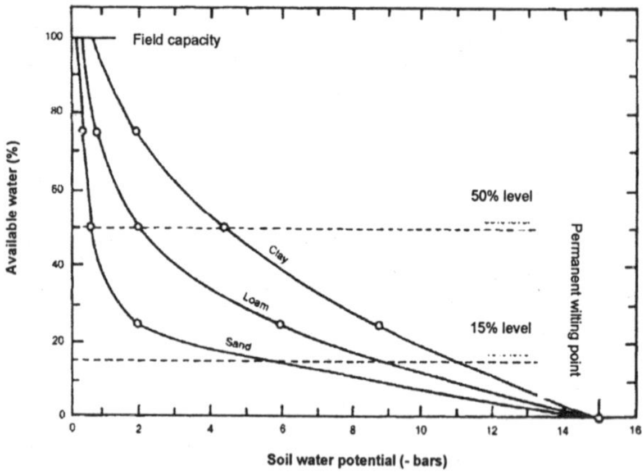

Figure 4.9. Water release curve for three soils.

Knowledge of the capacity of soils to retain available irrigation water is essential for efficient irrigation. If the farmer applies more water than the root-zone soil reservoir can retain at a single irrigation, the excess is wasted. If less is applied than the soil will retain, the plants may wilt from lack of water before the next irrigation unless water is applied more frequently. Water losses which result from deep percolation below the root zone of crops, cannot be seen. Losses can be determined or approximated by subtracting the storage capacity of the various soils from the amount of water applied in single irrigations, less the runoff.

WATER FLOW IN PLANTS

The water relations of plants are affected by soil factors like soil moisture, salinity, impeded drainage etc; by plant factors like crop species, root depths, drought tolerance etc., and weather factors like temperature, wind, rainfall etc. The plant essentially integrates its soil water and atmospheric environments and reflects the prevailing conditions in growth processes.

The plant-water relationship follows in three stages viz. absorption, translocation and transpiration. Plant roots absorb water through roots and transfer to the conducting tissues of the leaves. Water is lost from the plant due to transpiration through small cavities in the leaves called stomata.

Water moves in the soil-plant-atmosphere continuum in response to differences in the potential energy of water in the system. Transpiration causes a lower water potential in the plant shoot and root system than in the bulk soil; consequently, soil water moves into the root system along this potential gradient. Water first enters the root system through epidermal cells in contact with the moist soil, then in turn through cortical cells, the endodermis, pericycle cells, and finally to the xylem that transports the water to the aerial plant parts. The intensity of root development and physical contact between the root and soil are important physical considerations. When the upper part of the root zone becomes comparatively dry and water is available in the lower zone, the uptake of water per unit volume of soil has been observed to be proportional to the rooting density. Thus, the distribution of roots that varies with crop species and soil physical properties becomes an important management concern. The approximate magnitude of water potential in the soil-plant-atmosphere system is given in Table 4.3.

The difference in the total water potential in the soil-atmosphere system generates driving force for water movement from the soil through the plant to the atmosphere. The quantity of water used by plants for metabolic processes are very low compared to the water lost as transpiration.

Table 4.3. Water potential in the soil-atmosphere system.

Component	Water potential (bars)
Soil	- 0.1 to -20
Plant	-5.0 to –50
Atmosphere	-1000 to -2000

The presence of salts in the soil water solution must be considered when evaluating available water. The plant root contains a semi-permeable membrane that allows water to pass but not most of the salt. Therefore, the main effect of soluble salts on plants is osmotic with high salt levels making it difficult for the plant to obtain enough water from the soil solution to meet transpirational demand. Cell enlargement is affected initially, and plants exhibit the typical colour changes associated with water stress.

Soil-Water Extraction Depth

This is the soil depth used to determine the effective region of water uptake by plants. It is not necessarily the maximum rooting depth, especially for plants that have a long taproot. It is the depth to which an average mature plant can actively extract an appreciable amount of soil water. Because of the many factors that influence root development and proliferation, the effective depth must be determined for a specific location.

The root length density (length of roots per unit volume of soil) is generally greatest near the soil surface and declines with increasing depth to the maximum depth to which roots are observed for a given crop species. This general trend is illustrated graphically in Figure 4.10. Extraction of water is most rapid in the zone of greatest root concentration and under the most favourable conditions of temperature and aeration. Because water also evaporates from the upper few inches of soil, water is withdrawn rapidly from the top part of the soil profile. Soils normally show a more rapid loss of water at shallower depths until the potential becomes low enough to be rate limiting. Basic water-extraction curves, based on quarters of the root zone depth, indicate that almost all plants growing in soil that is uniform and adequately supplied with water, have similar moisture-extraction patterns.

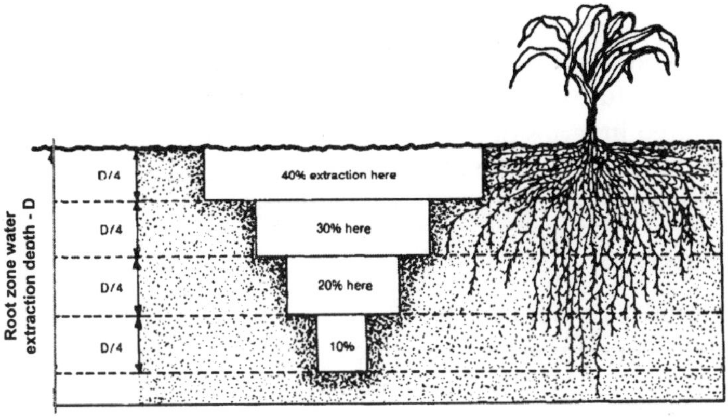

Figure 4.10. Average water extraction pattern of plants growing in a soil.

Figure 4.10 shows that about 40 percent of the extracted water comes from the upper quarter of the root zone, 30 percent from the second quarter, 20 percent from the third quarter, and 10 percent from the bottom quarter. Values for comparative crops are normally within 10 percent of this range. In non-uniform soils, the amount of soil water for crop growth may be determined by the soil layer that has the lowest soil water retention capability. For example, a topsoil layer with a low water retention capacity may be rapidly depleted following an irrigation or rain. Even though, soil water may be adequate at the lower depths, water stress could be experienced in the early stages of plant development if the root system is not yet fully established. The normal extraction pattern for a given crop will change when restrictive barriers are encountered. Also, if the water level in the upper soil layers is allowed to remain excessively dry, larger than normal amounts of water will be supplied by the lower soil layers.

EVAPOTRANSPIRATION

Water enters the soil through pores, cracks, worm and decayed root holes, and through cavities introduced by tillage. Infiltrated water may evaporate from the soil surface or may be transpired by the plants or may percolate downward beyond the plant roots and contribute to the groundwater.

Evapotranspiration (ET) is the process by which water is moved from the surface of the earth to the atmosphere. It consists of the evaporation of liquid or solid water from soil and plant surfaces, plus water that transpires through plant tissues. ET is measured in terms of inches or cms of water per day. ET varies with the plant, the climate, the level of soil moisture, and plant condition (fertilizer/pest/disease stress).

Potential ET

Potential ET refers to the maximum ET rate determined by climatic conditions for a specific crop at a specific location at a specific time. Climatic conditions largely determine the potential ET. Various methods, based on meteorological factors, have been developed by researchers to predict the potential ET rates. Solar radiation is the main factor that determines the ET rate; but air temperature, humidity, and wind speed also have an effect.

Direct measurement of ET rates is laborious, time consuming, and requires considerable instrumentation. Therefore, the measurement of climatic factors is most often used to estimate ET based on an equation or model that relates the climatic factors to the ET rate. ET models usually estimate the potential ET of a reference crop such as grass or alfalfa. Conversion factors called "crop coefficients" are used to relate the reference crop ET to the actual crop ET.

Some ET models, such as the pan evaporation and Modified Blaney-Criddle models, relate climatic factors directly to the crop rather than to a reference crop. Crop coefficients are still required to adjust for the plant development stages because the crop transpiration rate is directly related to the canopy leaf area. As the canopy area increases, the crop coefficients must be increased. When soil water is deficient, the plant is not able to take up enough water to meet the evapotranspiration demand set by the climatic conditions and is said to be under stress. When plants are stressed from soil water deficiency, the actual crop evapotranspiration rate (ET) will be less than the potential evapotranspiration rate (ET and the actual crop yield will be less than the maximum yield).

Crop Evapotranspiration

Water is extracted from the soil by evaporation at the soil and plant surfaces (crop transpiration). The combination of the two is termed "crop evapotranspiration", or ETc. It has also been termed "consumptive use" or "crop water use". ETc is affected by many factors.

ETc varies with the type of plant and growth stage. Some plants just use less water than others. And obviously, a seedling is going to extract less water than a full grown plant. ETc varies with the climate. Up to a certain point, increasing temperature and wind increases ETc. There is usually a maximum ETc rate for any plant, beyond which it just stops transpiring due to stress. With Increase in humidity and cloud cover ETc decreases.

ETc varies with the amount of soil moisture in the effective root zone. The soil's water-holding forces increase as the soil moisture decreases. Thus, as the plant uses up soil moisture, it becomes harder and harder to extract more. At a certain point, which depends on the plant, the soil's texture and structure, and the root zone salinity, the ETc rate decreases. Below-normal, ETc rates cause stress on the crop. Depending on the crop and growth stage, more or less stress is desirable. Knowing the acceptable level of stress and knowing what level of soil moisture will cause this stress is an important function of modern crop management. Table 4.4 contains approximate ETc's of common crops.

Table 4.4. Approximate ETc at maturity of some crops.

CROP	ETc (cm/day)
Alfalfa	0.75
Almond	0.70
Corn	0.90
Cotton	0.70
Melon	0.50
Onion	0.50
Tomato	0.75
Sugar beet	0.75

Predicting Crop ET

Actual crop ET, in addition to climate, depends on soil factors and plant factors such as the degree of ground cover, plant leaf characteristics, and surface roughness of the crop canopy. Plant factors are characterized by the crop coefficient that varies during the growing season and according to the model used to estimate ET.

Estimating the actual ET of a growing crop from climatic observations requires the reference crop ET and the specific crop coefficient. The ET of a specific crop is calculated by:

ETc = (reference crop ET) x (specific crop coefficient)

Standard terminology has been established for referring to the various ET models and crop coefficients.

Several equations or models are available for estimating reference crop ET. The two reference crops used for estimating crop ET are grass and alfalfa. Grass is the reference crop most often used and is becoming the standard reference. Care must be used to avoid mixing methodologies and coefficients. The selection of the method to be used may be determined by the available information.

Crop water requirement is defined as the depth of water per unit soil area needed to meet the water loss from evapotranspiration (ETcrop or ETc) of a disease-free crop growing in a large field under non-restricting soil conditions, including soil water and fertility, and achieving full production potential under the given growing environment. When these conditions are met, the crop will produce at the potential or maximum yield and transpire at the maximum rate (ET).

WATER DEPLETION

Growth of most agricultural crops is favoured by a soil water content that is high enough to encourage crop growth and development, but not so high that aeration becomes restrictive. These concepts are illustrated in Figure 4.11. If soil water is plant-extracted to levels approaching the PWP, water is held so tenaciously by the soil that plants can no longer obtain sufficient water to meet the potential for transpiration. Transpiration is restricted and yield losses take place. Excessive filling of the soil pore space with water excludes sufficient air to meet plant oxygen requirements, and yields are again reduced. Plant species vary in their tolerance to either deficits or excesses. Water management programs must reflect individual crop characteristics.

A critical water level varies with the soil as well as with the crop. The water content of the sand, however, is almost down to the wilting point. More energy is required for a plant to extract water from the clay at the 15 percent level than from the sand at that level, but more water is available in the clay soil which provides a greater safety factor. To provide a reasonable safety factor, the lower limit of water depletion in the sandy soil must be higher than 15 percent for most crops. To illustrate this point, suppose the sand, loam, and clay soils hold 0.7, 1.4, and 2.4 inches of plant available water, respectively, per foot of soil depth at field capacity or 100 percent available water. At 15 percent available water remaining, there are 0.10, 0.21, and 0.36 inch per foot of soil depth for the sand, loam, and clay soils. For a root profile depth of five feet, the sandy soil only has a total of 0.5 inch of available water remaining.

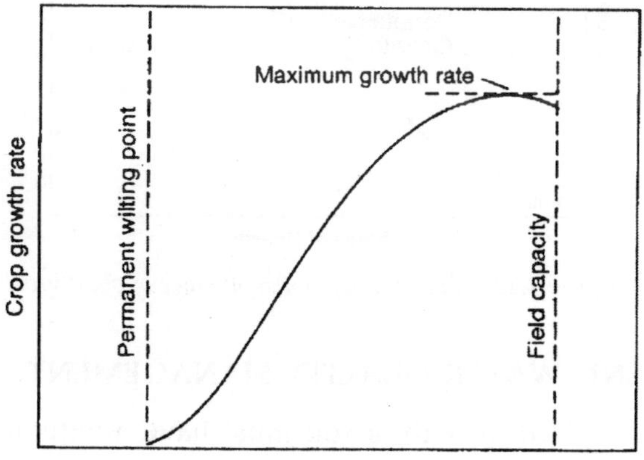

Figure 4.11. *Relationship between soil-water retention and crop growth.*

USE OF WATER BY PLANTS

A plant has different water needs at different stages of growth. While a plant is young, it requires less water than when it is in the reproductive stage. When the plant approaches maturity, its water need drops (Fig. 4.12). Curves have been developed that show the daily water needs for most types of crops.

A plant's root system must provide a negative tension (pressure) to extract water from the ground. This tension must be equivalent to the tension that holds the water in the soil. For example, if the water in the soil is at 0.3 bars (around field capacity), the plant must provide at least 0.3 bars of negative tension to pull the water from the ground. At the wilting point, the maximum negative tension that a plant can provide is balanced by the soil water tension. At this point, the plant can no longer extract water from the soil and will be permanently stressed. There are several factors that determine when, where, and how much water a plant will use. These factors include daily plant water need as influenced by climatic conditions and stage of growth, plant root depth, and soil and water quality.

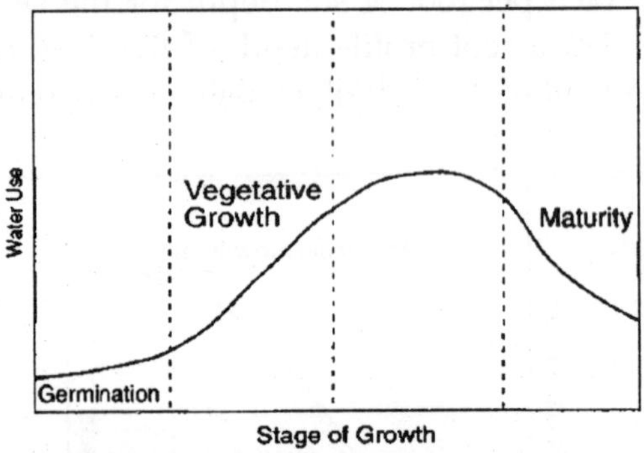

Figure 4.12. Water needs at different stage of plant growth.

SOIL AND WATER QUALITY MANAGEMENT

For good plant growth, a soil must have adequate room for water and air movement, and for root growth. A soil's structure can be altered by certain soil management practices. For example, excessive tillage can break apart aggregated soil and excessive traffic can cause compaction. Both of these practices reduce the amount of pore space in the soil, and thus reduce the availability of water and air, and reduce the room for root development.

The quality of the water is also important to plant development. Irrigation water with a high content of soluble salt is not as available to the plant, so a higher soil water content must be maintained in order to have water available to the plant. Increasing salt content of the water reduces the potential to move water from the soil to the roots. Some additional water would also be needed to leach the salt below the crop root zone to prevent build-up in the soil. Poor quality water can affect soil structure.

SUMMARY

Basic knowledge of soil-plant-water relationships make it possible to better manage and conserve water. Some of the important factors to remember include:

a) Soil water holding capacity varies with soil texture. It is high for medium and fine-textured soils but low for sandy soils.

b) Plant roots can use only available soil water i.e. the stored water between field capacity and permanent wilting point. However, as a general rule, plant growth and yields can be reduced if soil water in the root zone remains below 50 percent of the water holding capacity for a long period of time, especially during critical stages of growth.

c) Although, plant roots may grow to deep depths, most of the water and nutrients are taken from the upper half of the root zone. Plant stress and yield loss can occur even with adequate water in the lower half of the root zone.

d) Poor irrigation water quality can reduce the plant's ability to take up water and can affect soil structure.

The relationship between crop, climate, water, and soil is complex and involves many processes. The processes can be explained somewhat simplistically through a series of concepts and relationships established through research.

5

Problem Soils

Some soils are extremely productive while others are not so productive. Each soil has a set of unique characteristics which distinguishes it from other soils. These characteristics determine the potential productivity of the soil. Because of how they have been managed for agricultural production and otherwise changed by man's activities, some of these soils are less productive. Two of the most common causes of productivity loss are the development of acid and saline (including saline-alkali and alkali) conditions. These soils develop under different conditions and their treatment and management is also different. They are often recognized as problem soils because they do not respond to normal management.

ACID SOILS

Soil acidity is a crop production problem of increasing concern. Soil acidity exceeding a particular limit is injurious to plant growth. The availability of certain nutrients, particularly phosphorus, calcium and magnesium becomes low with increasing acidity. In acid soils ions of aluminium, iron, manganese and copper may be found in the dissolved form in quantities sufficient to become toxic. Similarly most of the desirable soil microbiological processes such as the activities of the *Azotobacter* and nodule forming bacteria of legumes are adversely affected as the acidity increases. The satisfactory granulation of soils also becomes difficult to achieve. It is therefore necessary to correct soil acidity for cultivating such soils profitably.

Causes of Soil Acidity

The four major causes for soils to become acid are listed below:

1. Acidic parent material
2. Rainfall and leaching
3. Organic matter decay
4. Harvest of high yielding crops

The above causes of soil acidity are most easily understood when we consider that a soil is acid when there is an abundance of acidic cations, like hydrogen (H^+) and aluminium (Al^{3+}) present compared to the alkaline cations like calcium (Ca^{2+}), magnesium (Mg^{2+}), potassium (K^+), and sodium (Na^+).

Parent material. Due to differences in chemical composition of parent materials, soils will become acidic after different lengths of time. Thus, soils that developed from granite material are likely to be more acidic than soils developed from calcareous shale or limestone.

Rainfall and leaching. Rainfall is most effective in causing soils to become acidic if a lot of water moves through the soil rapidly. Excessive rainfall is an effective agent for removing basic cations. Sandy soils are often the first to become acidic because water percolates rapidly, and sandy soils contain only a small reservoir (buffer capacity) of bases due to low clay and organic matter contents. Since the effect of rainfall on acid soil development is very slow, it may take hundreds of years for new parent material to become acidic even under high rainfall.

Organic matter decay. Decaying organic matter produces H^+ which is responsible for acidity. The carbon dioxide (CO_2) produced by decaying organic matter reacts with water in the soil to form the weak acid called carbonic acid. This is the same acid that develops when CO_2 in the atmosphere reacts with rain to form acid rain. Several organic acids are also produced by decaying organic matter, but they are also weak acids. Like rainfall, the contribution to acid soil development by decaying organic matter is generally very small, and it would only be the accumulated effects of many years that might ever be measured in a field.

Crop production. Harvesting of crops has its effect on soil acidity development because crops absorb lime-like elements, as cations, for their nutrition. When these crops are harvested and the yield is removed from the field, some of the basic material responsible for counteracting the acidity developed by other processes is lost, and the net effect is increased soil acidity. Increasing crop yields will cause greater amounts of basic material to be removed. Grain contains less basic materials than leaves or stems. For this reason, soil acidity will develop faster under continuous wheat pasture than when only grain is harvested. High yielding forages, such as Bermuda grass or Alfalfa, can cause soil acidity to develop faster than with other crops.

The use of fertilizers, especially those supplying nitrogen, has often been blamed as a cause of soil acidity. Although, acidity is produced when ammonium containing materials are transformed to nitrate in the soil, this is countered by other reactions and the final crop removal of nitrogen in a form similar to that in the fertilizer. The effect of nitrogen fertilizers has been to increase yields and thus increase the removal of bases.

Acid Soil Reaction

Knowing the soil pH helps identify the kinds of chemical reactions that are likely to be taking place in soils. Generally, the most important reactions, from the standpoint of crop production, are those dealing with solubilities of compounds or materials in soils. In this regard, we are most concerned with the effects of pH on the availability of toxic elements and nutrient elements.

Toxic elements like aluminium (Al) and manganese (Mn) are the major causes for crop failure in acid soils. These elements are a problem in acid soils because they are more soluble (available for plant uptake) at low pH. In other words, more of the solid form of these elements will dissolve in water when the pH is very low. There is always a lot of solid Al present in soils because it is a part of most clay particles.

Element toxicities. When soil pH is above about 5.5, Al in soils remains in a solid combination with other elements and is not harmful to plants. As pH drops below 5.5, Al containing materials begin to dissolve. Because of its nature as a trivalent cation (Al^{3+}), the amount of dissolved Al is 1000 times greater at pH 4.5 than at 5.5, and 1000 times greater at 3.5 than at 4.5. For this reason, some crops may seem to do very well, but then fail completely with just a small change in soil pH. Wheat, for example, may do well even at pH 5.0, but usually will fail completely at a pH of 4.0.

The relationship between pH and dissolved Mn in the soil is similar to that described for Al, except that manganese (Mn^{2+}) only increases 100 fold when the pH drops from 5.0 to 4.0.

Toxic levels of aluminium harm the crop by "root pruning". That is, a small amount of Al in the soil solution in excess of what is normal causes the roots of most plants to either deteriorate or stop growing. As a result, the plants are unable to normally absorb water and nutrients, appear stunted and exhibit nutrient deficiency symptoms, especially those for phosphorus. The final effect is either complete crop failure or significant yield loss. Often times the field will appear to be under greater stress from pests, such as weeds, because of the poor crop conditions.

Toxic levels of Mn interfere with normal growth processes in above ground plant parts. This usually results in stunted, discoloured growth and poor yields.

Soil buffer capacity and buffer index. Although, crops remove large quantities of lime-like materials that are harvested each year, the soil pH usually does not change noticeably from one season to the next. Because soil pH does not change quickly, it is said to be buffered. Buffer means the resistance to change of pH.

There are several reasons why soils have this buffer ability or buffer capacity. Soils commonly contain free calcium carbonate (lime). The term *caliche* is used to describe layers of soil material cemented by accumulated calcium carbonate. These accumulations provide a huge reserve of lime that will maintain soil pH in the alkaline range (above pH 7) for generations, perhaps centuries, even under the most productive agricultural systems.

A second contribution to the buffering capacity of soils is the release of basic chemical elements from normal chemical weathering of soil minerals. This is a very slow process that occurs whenever water is added to soil. The effect is influenced by the type of minerals in the soil, the amount and frequency of water addition, and soil temperature.

The most important source of buffer capacity in acid soils (no "free" lime present) is exchangeable cations. These are the lime-like chemical elements (mostly calcium) that are adsorbed on the surface of soil particles. These adsorbed basic materials act like a large reservoir that continually replenishes basic materials in the soil solution when they are removed by a crop or neutralized by acid. Figure 5.1 illustrates this and the relationship between soil pH and buffer capacity.

As crops remove bases from soil water in the reservoir on the right (Fig. 5.1), bases from the large reservoir of soil solids (clay and humus) on the left move to the soil solution and replenish the supply. Because of this relationship and the large reserve of bases from soil solids, the pH does not change much from month to month or even year to year. Also since the large reservoir on the left is shaped like a pyramid, pH can often be changed more easily by liming at pH near 6.0 than in the very acid pH 4.5 to 5.5 range.

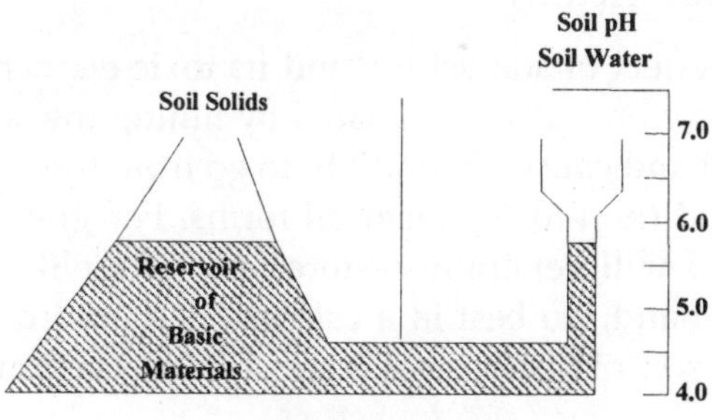

*Figure 5.1. **The relationship of basic materials in soil solids to pH of the soil solution.***

Figure 5.2 shows the influence of soil organic matter and texture on buffer capacity. Both soils have a pH of 4.3, and are too acidic for efficient crop production. In order to provide a more favourable pH, the soils must each be limed. The amount of lime required will depend on the size of the large reservoirs and how base depleted they may be.

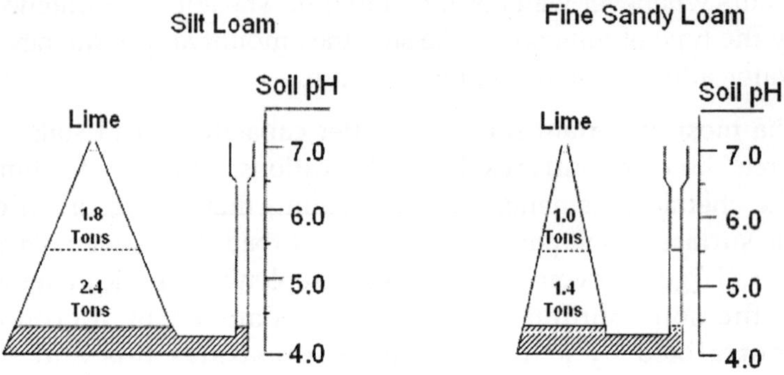

Figure 5.2 ***Reservoirs of soil solids in clayey vs. sandy soil.***

From these diagrams, it is easy to understand why it takes much more lime to raise the pH of a clay soil with its large reservoir than it does for a sandy soil and its small reservoir. Also, because the reservoir of sandy soil is small, if acidifying conditions are equal, sandy soil will tend to become acid more rapidly and need to be limed more frequently than clayey soil.

Correcting Soil Acidity

The adverse effect of soil acidity and its toxic elements is most easily (and economically) eliminated by liming the soil. Liming raises soil pH and causes Al and Mn to go from the soil solution back into solid (non-toxic) chemical forms. For grasses, raising soil pH to 5.5 will generally restore normal yields. Legumes, on the other hand, do best in a calcium rich environment and often need a soil pH between 6.5 and 7.0 for maximum yields.

Table 5.1. Common pH preference of field crops.

Crops	pH Range
	Legumes
Field bean	5.5 - 7.0
Berseem	6.0 - 7.5
Groundnut	5.3 - 6.6
Soybean	5.5 - 7.0
Pea, Lentil, Gram etc.	5.5 - 7.0
	Non-Legumes
Cereals	6.0 - 7.5
Maize	5.0 - 6.5
Millets	4.0 - 6.0
Rice	6.0 - 7.5
Sorghum	6.0 - 7.5
Wheat, Barley, Oats	5.0 - 7.7
Sugarcane	6.0 - 7.5
Cotton	5.0 - 6.5
Potato	5.0 - 5.0
Tea	4.0 - 6.0

Soil pH in the range of 6.0 to 7.0 is also desirable from the standpoint of optimum nutrient availability. The common pH preference of some crops is given in Table 5.1. However, the most common nutrient deficiencies are for N, P and K, and availability of these elements will not be greatly changed by liming. Nutrients most affected by soil pH are iron and molybdenum. Iron deficiency is more likely to occur in non-acid (high pH) soils. Molybdenum deficiency may also occur in acid soils and could be corrected by liming.

Lime reactions. Soil acidity can only be corrected by neutralizing the acid present, which is done by adding a basic material. While there are many basic materials which can neutralize acids, most are too costly or difficult to manage. The most commonly used material is agricultural limestone. It is used because it is relatively inexpensive and easy to manage.

The reason limestone is easy to manage is because it is not very soluble, meaning it does not dissolve easily in water. For this reason, it is not very corrosive to equipment and more importantly, its pH at equilibrium (after it has dissolved as much as it can and there is still some lime left in the water) is only about 8.3. This latter aspect is very important because even if an excessive amount of lime is applied, a harmful effect on crop yields would generally not take place.

The reaction of lime, or calcium carbonate ($CaCO_3$), with an acid soil is illustrated by Figure 5.3.

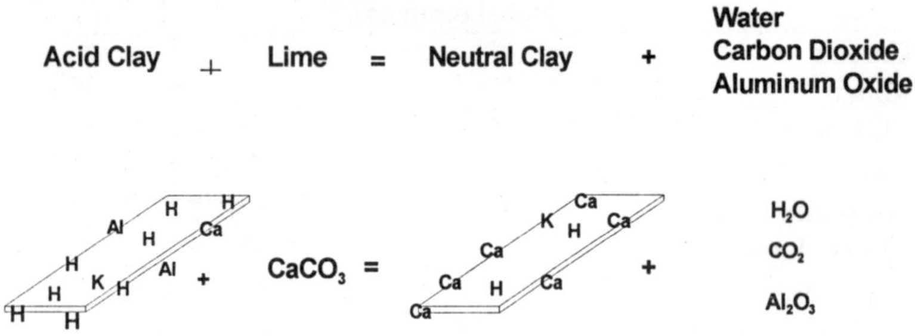

Figure 5.3. ***Illustration of how agriculture lime neutralizes soil acidity.***

This diagram shows that the acidity is on the surface of soil particles. As lime dissolves in the soil, calcium (Ca) from the lime moves to the surface of soil particles and replaces the acidity (H^+ and Al^{3+}). The acidity reacts with carbonate (CO_3) to form carbon dioxide (CO_2) and water (H_2O). The end result is a soil that is less acid.

Lime requirement

The lime requirement of a soil necessary to neutralize the total acidity may be defined as the amount of liming material that must be added to raise the pH usually in the range of 6.0 to 7.0. The pH measurements are widely used to estimate the lime requirements. The basis for this is that in acid soils a relationship exists between the pH and the percentage base saturation of the soil. Once this relationship is known, it is useful for estimating the lime requirement in the laboratory. A relationship between soil test pH values and expected wheat yield is illustrated in Figure 5.4. The yield at a given pH is expressed as relative yield. This term means the expected yield as a percentage of that possible if soil acidity is not a limiting factor. For example, if a 40 q yield is expected with no acidity problems then at a soil pH of 5.0 a relative yield of 85%, or 34 q (40 x 0.85), would be expected.

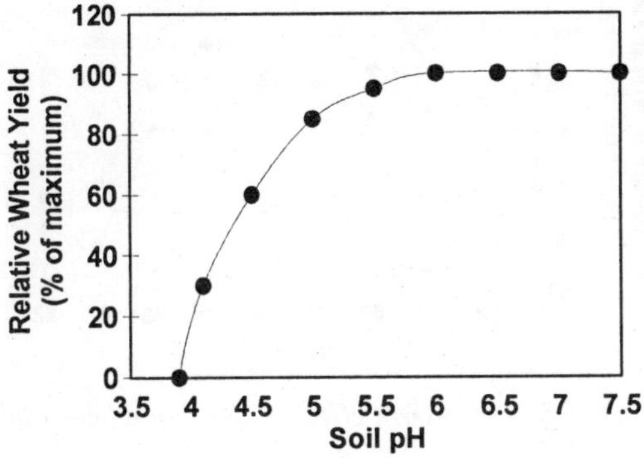

Figure 5.4. The effect of soil pH on wheat yield.

However, with this "lime requirement" under crops conditions the predicted pH is not generally attained. Hence a "liming factor" of 1.5 to 2 is commonly used to achieve the desired results i.e. the lime requirement, as determined in the laboratory, is multiplied by a factor of 1.5 to 2.

Several field research experiments have been conducted on various crops over the past 20 years to examine suitable liming materials and application rates. A common feature of all effective commercially available liming materials is that they contain a basic lime-like material such as calcium or magnesium carbonate. Since it is ultimately the material from which other basic materials are derived. Agriculture Lime is usually the lowest cost per ton of active ingredient (ECCE or Effective Calcium Carbonate Equivalent, finely ground pure $CaCO_3$ is defined to have an ECCE of 100).

Minimum amount. The amount of lime to apply for crop production depends on cultivation of crops alone or in rotation with a legume. For example, if wheat alone is grown year after year, it is necessary to only apply a rate of lime to raise the pH to above 5.5. If legumes are sometimes grown then soil pH should be raised to 6.5 or above. Figure 5.5. show that the yield of wheat greatly improved by a single application of lime.

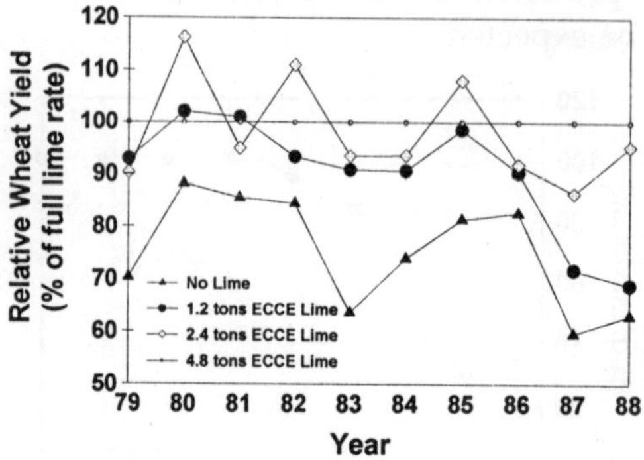

Figure 5.5. ***Long-term effect of lime on wheat yield.***

Calculating rates. Lime requirements are expressed in terms of ECCE. There are two components to the determination by the laboratory. First, the purity of the lime is determined chemically (purity factor). In this test fraction of $CaCO_3$ is analysed for the, or its equivalent, in the lime material. The second measure is a determination of how finely the lime particles are ground (fineness factor). The fineness factor is determined by weighing sieved portions of a lime sample. The factor is then calculated by taking ½ times the fraction (e.g. 0.90) of sample passing an 8 mesh sieve plus ½ times the fraction (e.g. 0.70) of sample passing a 60 mesh sieve. The fineness factor for these example values would be:

$$\frac{1}{2} \times 0.90 + \frac{1}{2} \times 0.70 = 0.80$$

The purity factor (a fraction) and the fineness factor (a fraction) are multiplied and then times 100 to obtain the ECCE value. If the purity factor is 0.90 (90% pure or equivalent calcium carbonate) then the ECCE would be (0.90 x 0.80) x 100, or 72%. The more $CaCO_3$ in the material and the finer the particle size, the greater the ECCE. Good quality lime will have an ECCE value above 50 percent. Because agriculture lime does not always have an ECCE of 100 percent, the amount required to provide a given amount of 100 percent ECCE must be calculated. The calculations to use are shown below:

$$\frac{\text{ECCE lime required}}{\text{\% ECCE}} \times 100 = \text{Agriculture Lime required}$$

For example, let us assume that the available Agriculture Lime is 72% ECCE and the soil test indicated a need for 1.5 tons ECCE to raise the soil pH to the desired level. The calculations would be:

$$\frac{1.5 \times 100}{72} = 2.1 \text{ tons of Agriculture Lime}$$

So, 2.1 tons per acre of the 72% ECCE lime would have to be applied in order to get the 1.5 tons of 100 percent ECCE lime required to do the job.

The approximate amount of limestone required to raise the pH to the neutral level for some soils is given in Table 5.2.

Table 5.2. Quantity of limestone required (tonnes/ha)

Soil	from pH 4.5	from pH 4.6	from pH 5.5
Warm humid plains			
sands and loamy sands	3.75	2.5	1.25
sandy loams	-	5.0	2.5
loams and silt loams	-	8.75	5.0
clay loams	-	12.5	7.5
Cool temperate hilly region	7.5	5.0	2.5
sands and loamy sands	-	7.5	5.0
sandy loams	-	11.25	7.5
loams and silt loams	-	15.0	8.75
clay loams			
Valley soils	22.5	17.5	11.5
organic, water logged			

Lime applications

Because lime does not dissolve easily in water, it must be treated similarly to fertilizers that supply the soil with immobile nutrients like phosphorus. Thus, for lime to be most effective in neutralizing soil acidity it must be thoroughly mixed with the soil. Since neutralization involves a reaction between soil particles and lime particles, the better lime is mixed with the soil, the more efficiently the acidity is neutralized. For this reason, wet materials which cannot be thoroughly mixed with the soil are often less effective. Similarly, pelleted lime particles are too large to mix well with small soil particles. Attempts to mix these materials with soil often result in soil acidity being neutralized only near the lime aggregates (or pellets), whereas acidity between aggregates remains unaffected. Once the proper rate has been determined and the lime has been spread to give a uniform application over the field, it is best to incorporate it with a light tillage operation such as disking. Disking can be followed by ploughing, but care should be taken not to plough too deeply as the lime will be diluted by subsoil and be less effective. Lime rates are calculated on the basis of neutralizing the top six inches (15 cm) of soil.

Since the lime reaction involves water, the effect of lime will be very slow in dry soil. Even when everything is done correctly and the soil is moist, it often takes a year or more for a measurable change in soil pH to occur. For this reason, liming for field crop production should be done as soon after harvest as possible. However, when the soil pH is extremely low, sufficient change may occur in just a few weeks and make the difference between being able to establish a crop and having a failure.

Periodic soil testing will identify lime needs early enough to maintain desirable soil pH by unincorporated broadcast application.

Liming materials

The most common and most effective liming material continues to be ground agriculture lime. It is generally be powdery with only a small percentage of coarse (sand size) particles, and have an ECCE of 50% or greater.

"Liquid Lime" is a formulation of high quality agriculture lime (usually ECCE is above 90%) with water and enough clay to keep the lime in suspension. The amount of water added may range from 35 to 50%. Care should be taken to make sure that the added water is not being charged for, as if it is high quality lime. When 90% ECCE lime is mixed 50% (weight to weight) with water, the resulting product is only 45% ECCE lime (90% x 0.50 = 45%). The fact that it is suspended in water does not increase its effectiveness. On the contrary, wet lime may not mix as easily with soil and therefore, its neutralizing effectiveness may be less than an equal amount of dry ECCE agriculture lime.

Similarly, "water treatment lime" may not be as effective as an equal rate of agriculture lime. This material is a waste product from water treatment plants. Although, it has a high ECCE, it is often wet when applied and a good mixture with soil is difficult to obtain. Too often, large chunks or globs remain mixed with the soil and only the acid soil next to the chunk of lime is neutralized, leaving large areas of soil between chunks that remain acid.

Pelleted lime is finely ground lime that is pressed into pellets. Until the pellets physically break up and the fragments of powder size lime become thoroughly mixed with soil, these too are limited in neutralizing soil acidity. Pellets, liquid lime, and water treatment lime can be spread or applied without dust common to good agriculture lime. Although easily visible, air-born dust associated with agriculture lime application represents only a small fraction of the total applied, and loss from the field should not be significant.

Finally, sometimes coarse "low grade" lime is in abundance and can be purchased at a very low cost. These lime materials are basic slag, press mud, etc. This cheap lime is too coarse to have a reasonable ECCE.

Reducing Metal Toxicity

Fertilizer reactions. Phosphate in the soil has long been known to be less available to crops in some extremely acid soils because it reacts with aluminium and/or manganese, which are more available in acid soils. When phosphate reacts with these metals, the compound formed is a very insoluble solid (such as aluminium phosphate). As a result, not only is the phosphate unavailable, but the aluminium and manganese are also unavailable. For these reasons, when phosphate fertilizers are banded with the seed at planting time, the harmful effects of toxic aluminium and manganese are greatly reduced and near normal yields may be obtained. Figure 5.6 illustrate the benefit of this practice for both grain and forage production.

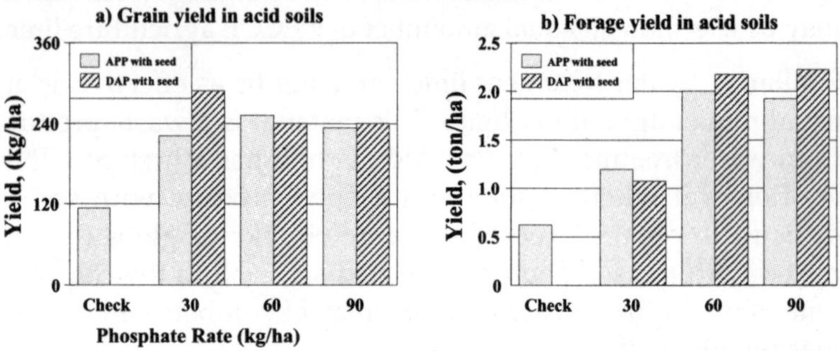

Figure 5.6. Responses of wheat grain and forage yields to seed-applied phosphate fertilizers (APP: ammonium polyphosphate; DAP: diammonium phosphate) in a strongly acidic soil.

Phosphate materials and rates. Figure 5.6 also shows that a higher rate of phosphate may be needed in order to get maximum benefits for fall forage production. It is especially important to use the higher rate for forage production on soil that has a pH below 4.5. The use of phosphate fertilizer in this way does not change soil pH. Also, within a few months after all the phosphate has been "used up", more aluminium and manganese may become available. While this may not affect the developed crop, it will affect the next crop in the seedling stage. As a result, phosphate fertilizer must be applied each year, whereas lime only needs to be applied every three to five years. On the other hand, build up of soil test P above crop needs may lead to increased P in the runoff.

When to use phosphate. As stated earlier, acid soil is best neutralized by adding agriculture lime. However, seed-applied phosphate (either ammonium polyphosphate or diammonium phosphate) should be considered for acid soils when the soil acidity problem is discovered too late for lime application in a given season and the soil has a low soil test value for phosphorus.

It is important to remember that this use of phosphate fertilizer is very different from normal. Banding phosphate on acid soils can increase yields even when the phosphate soil test value is very high (>65); not because more phosphate is provided to the plant, but because metal toxicity is reduced. Also, it is important to remember that the soil continues to become more acid with time. Eventually lime must be added to the soil to neutralize acidity.

SALINE AND ALKALI SOIL

Two other problem soils are salty (saline) soils and alkali or sodic (slick-spot) soils. A third problem soil often develops from slick-spots when they are poorly managed. This is the saline-alkali soil which results when slick-spots become salty.

Although, all problem soils may be identified with poor crop production, these soils have other similarities and differences which are important to know before attempting to improve or reclaim them.

Saline soils are soils which contain at least 2600 ppm soluble salts in the solution from a soil saturated with water. The salt content is estimated by laboratory measurement of how well the soil water conducts electricity, and saline soils are those with an electrical conductivity (EC) of 4000 micromohs/cm (about 2600 ppm). This level of salts is great enough to reduce production of salt-sensitive crops. Normal, productive agricultural soils commonly have electrical conductivity values below 1000.

Alkali soils are soils which contain enough sodium (Na) to cause 15% of the cation exchange sites to be occupied by Na (exchangeable sodium percentage, ESP). Sodium in the soil, prevents clay particles (and other very small, colloidal sized particles such as humus) from coming together and forming large soil aggregates. When soils contain 15% or more of exchangeable sodium most of the clay and humus particles are unattached or dispersed. These soils commonly have a pH of 8.5 or above (alkali). Some soils become dispersed even when the exchangeable sodium is as low as 7 percent. Productive agricultural soils often have less than 1 percent exchangeable sodium. Soils can be classified into 4 groups based on the EC and ESP of saturated paste extract. They are illustrated in Fig. 5.7.

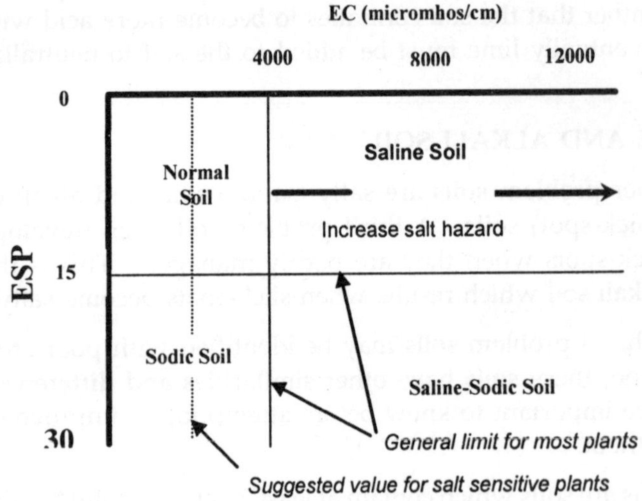

Figure 5.7. General classification of salt affected soils.

Characteristics of Saline Soils

Small growing areas affected. Naturally developed saline soils usually represent only small areas of a field. Often these are low lying parts of the field which may have poor internal soil drainage. Other small areas occur on slopes where erosion has exposed saline or alkali subsoil. Because low areas are frequently wet when the rest of the field is dry enough for cultivation, these small areas frequently are cultivated when the soil is too wet. This results in the soil becoming compacted in and around the area. Water does not move easily through the compacted soil so more water evaporates, leaving salts from the water to accumulate. As a result, the affected area increases with time.

Poor yield. Crop production is usually less than normal in salt affected areas. Yield reduction is greatest in years of less than normal rainfall or when water stress has been a yield limiting factor. Salts 'tie-up' much of the water in the soil and prevent plants from absorbing it. Seedlings are the most sensitive to water stress and crop stand is reduced because of seedling death and poor yield results.

White surface crust. As water evaporates from saline soils, salts which were in the water are left behind to accumulate on the soil surface. Salts are light coloured and when accumulation has continued for several days they form a very thin white film on the soil surface. During hot, dry weather, the light film will show up first along edges of the salt problem areas. The centre of these areas usually has the most salt and will dry out last.

Good soil tilth. Saline soils generally have excellent physical conditions throughout the tillage depth. This is caused by salts effectively neutralizing the negative charge of clay particles, allowing them to attach to one another. When these soils are not too wet, the soil is friable, mellow, and easily tilled. The appearance after tillage is that of a very productive soil.

High soil fertility. Soil which has been saline for several years will usually be very fertile, and high N, P, and K soil test values are often a clue of a problem salty soil. These nutrients build up in salty areas when there is little crop nutrient removal and the area is fertilized each year. Soil pH does not change in relation to salt content and it cannot be used as an indicator.

Characteristics of Alkali Soils

Except as noted, alkali soils have characteristics similar to saline soils. For this reason, one problem soil may be confused with another. Their differences, however, are important to note as they relate to correcting the problem soils.

Poor soil tilth. The excess sodium in alkali soils does not allow soil particles to easily attach to one another. As a result, alkali soil is dispersed and not friable and mellow like saline soil. Instead, alkali soil is greasy when wet ('slick-spot'), especially if it is fine textured, and often very hard when dry. This poor physical condition makes these soils difficult to manage. They are often either too wet or too dry for tillage. Poor seed germination and stand establishment are common because good seedbed preparation is seldom accomplished. As a result, yields are usually lower than the rest of the field and fertility may build up.

Dark or light coloured surface. Soil colloids which are floating in the soil water are left as a thin film on the surface after water evaporates. The surface colour will be darker than the rest of the field (black-alkali) when the particles are mainly humus since humic acid dissolves in alkali solution and lighter (white-alkali) when the particles are mainly clay and salts. The salts show up as a film when the surface dries.

Droughty but pond water. Large pores or channels in the soil which allow water entry and penetration become plugged with dispersed clay and humus. As a result, the subsoil may be very dry even though water is ponded on the surface. Plants that do become established often suffer water stress and may eventually die from lack of water and/or oxygen.

Reclamation

In many instances, saline soils and alkali soils can be reclaimed by following a definite series of management steps designed to leach or "wash out" the salts or sodium. The order and description of these steps follows.

Verify problem. The first step to solving the problem is clearly identifying it. This is best done by having the soil tested. Suspected areas should be sampled separate from the rest of the field. It is best to sample during a dry period of the growing season when affected areas of the field can easily be identified by poor crop growth. Samples should be taken at least one week from the last rain or irrigation and only the top 3-4 inches (8-10 cm) of soil should be sampled. Several small samples of the affected area should be combined in a plastic bucket and mixed to get a good sample. This test will identify the type and severity of the problem.

Identify cause. Whenever possible, it is important to find out what has caused the problem soil to develop. Knowing the cause can help in modifying the remaining reclamation practices and sometimes provide a clue as to how long it may take to complete the reclamation. The four most common causes of saline and alkali soils are:

a) naturally poor drainage
b) poor irrigation water
c) brine spills
d) exposure of saline or alkali subsoil due to erosion.

Poorly drained soils are simply soils which water does not easily penetrate. This condition may be a result of the soil having a high clay content, having a water table near the surface (within 3 m), or existing in a low lying area of the field. In the latter situation, normally adequate internal drainage may not be able to handle runoff from the surrounding area. In some instances, internal soil drainage is greatly reduced as a result of compacting the surface soil.

Use of poor quality irrigation water may cause problem soils to develop if special precautions are not taken. The problem develops most rapidly during extremely dry years when evaporation and the amount of irrigation are high. Internal soil drainage may be a contributing factor.

Problem soils sometimes develop 'overnight' when brine solutions associated with oil and gas well activities spill onto the soil. Depending on the amount of brine solution spilled and the size of the area, the problem may be slight or very severe.

Whenever the source of salt or sodium causing the problem is the result of addition from runoff, seeps, irrigation water or spilled brine, it is important to eliminate that source as soon as possible.

Improve internal soil drainage. There are no chemicals or soil amendments that can be added to the soil to "tie-up" or somehow inactivate soluble salts or sodium. Hence, the only way of lowering their concentration in the soil is to remove them. This can only be done by leaching (washing out) the salt or sodium downward out of the root zone. In order for this to happen, internal drainage must be good so water can easily pass through the soil.

There are a number of ways internal drainage can be improved. Most are expensive, but when the problem is severe, many will pay for themselves with time. Tile drains and open ditches are effective for removing subsoil water that accumulates due to a restrictive layer such as compacted clay or bed rock. Compacted soil layers near the surface can be broken up by subsoiling. This is effective only if done when the soil is dry enough to have a shattering effect and at best provides only temporary benefit.

Problem soils which have developed from use of poor irrigation water or brine spills may already have good internal soil drainage.

Add organic matter. Once internal drainage has been assured, the next important step is to improve water movement into the soil. Incorporating 50-75 tons per ha of organic matter into the top six inches (15 cm) creates large pores or channels for water to enter. Even rainfall from intense storms is more effective because there is less runoff. In addition to improving water movement into the soil, the large pores lessen the capillary or wick-like upward water movement during dry periods. Any coarse organic material such as barnyard manure, straw, rotted hay, or crop residue is suitable.

Add gypsum to sick-sots. Up to this point the reclamation practices are the same for both saline and alkali soils. In either situation, leaching is critical to remove salt or sodium. However, since high amounts of sodium absorbed to the soil are the cause of alkali problems, sodium must be loosened from the soil before it can be leached out. Gypsum is the most effective soil amendment for removing sodium from the soil particles. Gypsum is a slightly soluble salt of calcium sulphate. This means that gypsum will slowly react in the soil, but for a long time. The reaction is illustrated in Figure 5.8.

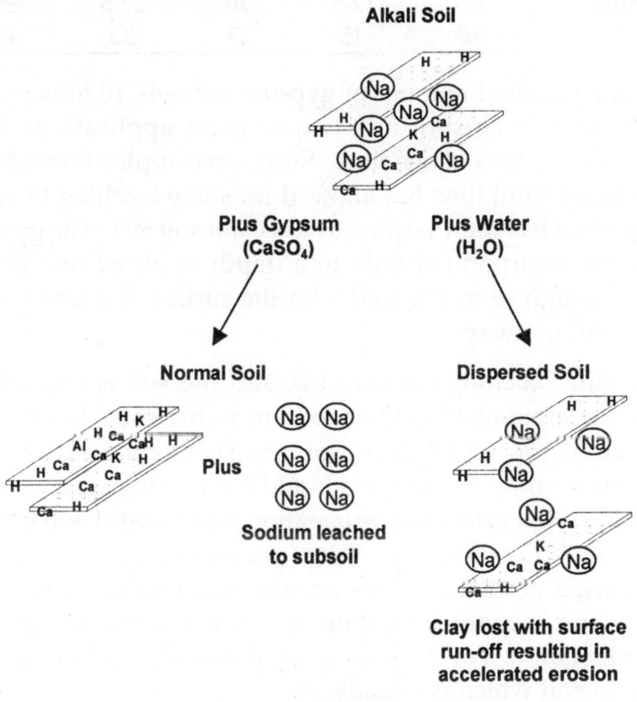

Figure 5.8 Alkali soil reacting with gypsum to form normal soil.

Gypsum applications are needed when the exchangeable sodium percentage (ESP) approaches 15 percent. Calcium ions (Ca^{2+}) in gypsum replace sodium ions (Na^+) on the colloids which results in improved soil conditions. The amount of gypsum required will vary widely depending upon the percentage of exchangeable sodium and the soil texture, as determined by the soil test. This relationship is shown in Table 5.3.

Table 5.3 Gypsum requirement in tons per ha as related to soil texture and sodium percentage.

Texture	Exchangeable Sodium Percentage				
	15	20	30	40	50
	gypsum (ton/ha)				
Coarse	5	7.5	12.5	7.5	22.5
Medium	7.5	12.5	20	27.5	35
Fine	10	15	25	35	45

When the required amount of gypsum exceeds 10 tons per ha, the rate should be split into two or more applications of no more than five tons at one time. Successive applications should not be made until time has allowed for some leaching to occur, and the need has been verified by a second soil test. The gypsum should be incorporated only to a depth of about one or two inches, enough to mix it well with the surface soil and keep it from blowing away.

Leach soil. Leaching (or washing out) the soil is essential to reduce the amount of salts or sodium in the soil. In order for this leaching process to occur, water must enter the soil in excess of what is used by growing crops and lost by evaporation. How fast and to what extent the reclamation is successful will depend on how much good quality water passes through the soil in a given period of time. The shorter the time interval over which excess water is applied, the more effective that amount of water is in reclamation. For this reason, rainfall is most effective when it falls on soil which is already wet.

Avoid deep tillage and establish cover. Once the leaching process has been started, deep tillage such as mouldboard ploughing should be avoided for several years to promote uninterrupted downward movement of the salts. Such tillage will bring salt back up to the soil surface, and leaching is then again required. As soon as the salt level in the soil is low enough, a salt tolerant crop such as barley or bermuda grass should be established on the problem area to provide a cover for as much of each growing season as possible. It is especially important to have the cover crop during midsummer when evaporation is high. Adequately fertilized bermuda grass does a good job of drying the soil. To minimize soil compaction, it should be cut for hay instead of pastured, make sure to keep heavy equipment off the area when it is wet.

Some problem areas may be too salty to establish a cover crop until some salts have been leached out. A cover crop can be established when there is no longer a white salty film on the soil surface, following a week or two of dry weather, or when weeds begin to grow.

Wait. The final step in reclamation is simply to wait for the previous practices to work. Except for brine spills, these problem soils developed over a period of several years. Reclamation may not take as long, but, depending on how well reclamation practices can be carried out, may take one or more years.

Alternative to Drainage

Learn to live with it. The key to successful reclamation is good internal soil drainage. If salts or sodium cannot be leached out, the soil cannot be reclaimed by conventional methods. However, most soils have some internal soil drainage, and although drainage may not be good, over several years time it may be sufficient to lower the salt concentration to near normal. During this time, it will be important to practice some of the same steps outlined above. Especially important are the following:

1. Avoid excessive fertilization.
2. Avoid traffic on field when wet.
3. Apply gypsum to slick spots.
4. Establish cover crop.
5. Maintain high level of crop residue.

Depending on the severity of the problem, it may be necessary to select a different crop than has been grown in the past. A list of crops and their relative tolerance to salt is provided in Table 5.4.

Table 5.4. The relative salt tolerance of crops.

Tolerant	Moderately Tolerant	Sensitive
	In increasing order of tolerance	
	Field Crops	
7,800-10,400 ppm	3,900-7,800 ppm	2,600 ppm
Cotton	Sunflower	Field beans
Sugar beet	Corn	
Barley (grain)	Soybeans	
	Grain sorghum	
	Oats (grain)	
	Wheat (grain)	
	Rye (grain)	
	Vegetable Crops	
6,500-7,800 ppm	2,600-6,500 ppm	1,950-2,600 ppm
Spinach	Cucumber	Green beans
Asparagus	Squash	Celery
Kale	Peas	Radish
Garden beets	Onion	
	Carrot	
	Bell pepper	
	Sweet potato and yam	
	Potato	
	Sweet corn	
	Lettuce	
	Cauliflower	
	Cabbage	
	Broccoli	
	Tomato	
	Fruit Crops	
	Cantaloupe	Strawberry
	Grape	Peach
		Apricot
		Plum
		Apple
		Pear

	Forage Crops	
7,800-11,700 ppm	2,600-7,800 ppm	1,300-2,000 ppm
Wheatgrass	Smooth bromegrass	Red clover
Bermudagrass	Oats (hay)	White clover
Barley (hay)	Wheat (hay)	Peanuts
Saltgrass	Rye (hay)	
Rhodesgrass	Alfalfa	
	Sudangrass	
	Perennial ryegrass	
	Yellow sweetclover	
	White sweet clover	

*Salt tolerance values at which 50% yield reduction may be expected compared to non-saline conditions.

SUMMARY

Highly acidic and saline/alkaline soils both limit crop growth and should be ameliorated for optimum crop production. Strongly acid soils with pH less than 4.5 brings down the soil micro-flora activity and increases toxicity of elements like iron, copper and aluminum. When soluble salts and/or high exchangeable sodium accumulate on the surface of the soil beyond the tolerable limit of the crops, the plant growth is affected adversely and the soil becomes unfertile. The acid soils can be made fertile by increasing its pH and replacing H^+ and Al^{3+} ion by Ca^{2+} and to some extent by Mg^{2+}. The basic requirement in the reclamation of saline and alkaline soils is the removal of salts and exchangeable sodium.

Depending upon the texture of the soil and the extent of damage, the management of soil assumes a great importance. No one method of reclamation will give dividends. Combination of 2-3 methods followed by well planned programme will be the sure way of success in reclaiming the acid, saline and alkali soils. Costly amendments will bring about the improvement in a short period, while other methods will have to be pursued for few years to get a positive result.

6

Soil Quality

Clean water and clean air is essential to maintain a healthy environment, yet few people recognize that our well being also depends upon high quality soil. For modern societies, soil is more important than ever before because the soil resources of the world are finite, non-renewable over the human time frame, and liable to misuse and mismanagement.

Nutrient cycling, water regulation, and other soil functions are normal processes occurring in all ecosystems. From these functions many benefits come to humans, such as food production, water quality, and flood control, which have value economically or in improved quality of life. Many important benefits are long-term or go beyond the land being managed. Off-site benefits of soil relate to the following resource issues:

- water quality of streams, lakes, oceans, and groundwater
- air quality, especially particulates
- greenhouse gases, including carbon dioxide, methane, and nitrous oxide
- biodiversity
- water flow and flood control
- sustainability of land productivity, and
- *aesthetics*

People can increase or decrease the value of soil benefits because land-management choices affect soil functions. Thus, it is important to understand what benefits we derive from soil and their value so we can appreciate the importance of managing land in a way that maintains soil functions. Important soil function and benefit of soil value to human is presented in Table 6.1.

Table 6.1. Soil function and benefits to humans.

Soil Function	Benefit of value to humans	
	On-site	**Off- site**
Nutrient cycling	Delivery of nutrients to plants	Enhances water and air quality
	Carbon storage improves a variety of soil functions	Storage of N and C can reduce greenhouse gas emissions
Maintaining biodiversity and habitat	Supports the growth of crops, rangeland plants, and trees	Helps maintain genetic diversity
	May increase resistance and resilience to stress	Supports wild species and reduces extinction rates
	Reduces pesticide resistance	Improves aesthetics of landscape
Water relations	Provides erosion control	Provides flood and sedimentation control.
	Allows on-site water recharge of streams and ponds	Groundwater recharge
	Makes water available for plants and animals	
Filtering and buffering	Can maintain salt, metal and micronutrient levels within range tolerable to plants and animals	Improves water and air quality
Physical stability and support	Acts as a medium for plant growth	Stores archeological items
	Supports buildings and roads	Stores garbage
Multiple functions	Sustains productivity	Maintains or improves air and/or water quality

INHERENT AND DYNAMIC SOIL QUALITY

Soil quality means the capacity of a soil to function, within ecosystem and land use boundaries, to sustain productivity, maintain environmental quality, and promote plant and animal health. A definition that includes an inherent and a dynamic component of a soil. The former is an expression of the soil forming factors and latter refers to the condition of soil that is changeable in a short period of time by human impact, including agricultural management practices.

Inherent properties. It changes little, if at all, with land use or management practices (Fig 6.1). They may include soil texture, depth to bedrock, type of clay, CEC, and drainage class. These properties are established as soil formed over millennia. How soils form depends on the five soil-forming factors-

- climate (precipitation and temperature)
- topography (shape of the land)
- biota (native vegetation, animals, and microbes)
- parent material (geologic and organic precursors to the soil)
- time (time that parent material is subject to soil formation processes)

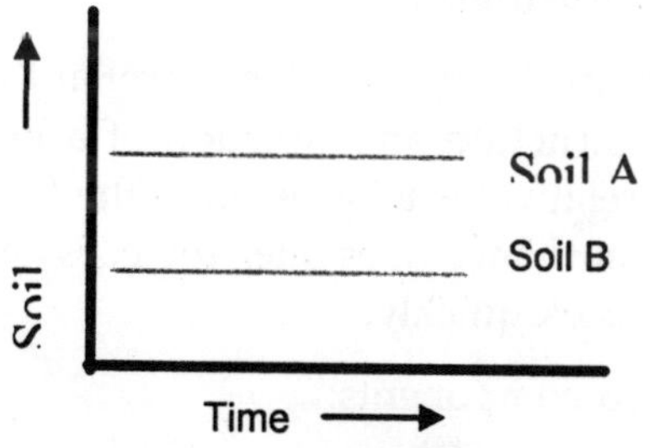

Figure 6.1. Inherent soil quality over time.

Dynamic properties. It can change over the course of months and years in response to land use or management practice changes (Fig 6.2). Dynamic properties include organic matter, soil structure, infiltration rate, bulk density, and water and nutrient holding capacity. Changes in dynamic properties depend both on land management practices and the inherent properties of the soil. For example, the organic matter levels in soil depend on tillage practices and the types of plants growing (management), but the total amount of organic matter is constrained by soil texture and climate (inherent features). Some properties, such as bulk density, may be considered inherent properties below 20-50 cm, but are dynamic properties near the surface.

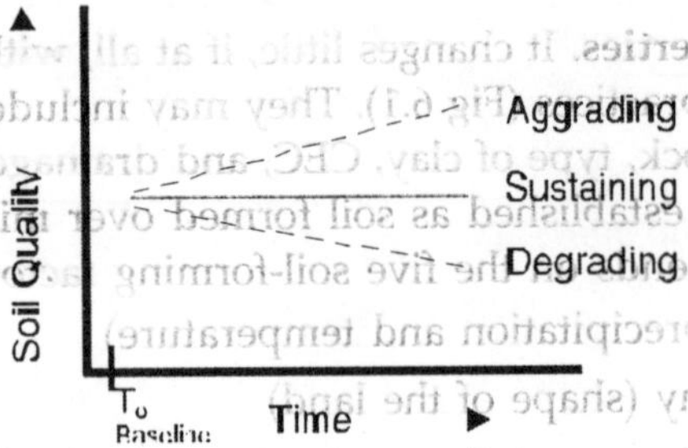

Figure 6.2. Dynamic soil quality changes over time.

ECOSYSTEM STABILITY

Resistance and Resilience

Ecosystem stability is an important corollary of sustainability. Over time, the structure and function of a healthy ecosystem should remain relatively stable, even in the face of disturbance. If a stress or disturbance does alter the ecosystem, it should be able to bounce back quickly.

Stability has two components:

Resistance- the ability of the ecosystem to continue to function without change when stressed by disturbance.

Resilience- the ability of the ecosystem to recover after disturbance.

Factors affecting stability:

- Disturbance frequency and intensity (how often and what kind of tillage).
- Species diversity (intercropping or rotations), interactions (competition for water and nutrients from weed species), and life history strategies (do the species grow fast and produce many seeds or slow with few seeds).
- Trophic complexity (how many functions are represented), redundancy (how many populations perform each function), food web structure (how do all of these groups interact).
- Rate of nutrient or energy flux (how fast are nutrients and energy moving in and out of the system).

Disturbance and Stability

An ecosystem disturbance can be natural or human induced stress. An example of a natural disturbance is a hurricane or a tornado. An example of a human-induced or anthropogenic disturbance is tillage or pesticide application.

Redundancy in ecosystem structure and function often infers stability on a system. For instance, if there is more than one (redundant) population of microbes that convert ammonium to nitrate and a disturbance wipes out one population, that function (nitrification) will continue to be performed by the remaining populations.

Because agro-ecosystems have reduced structural and functional diversity, they have less resilience than natural systems. The expected outputs from the system (yield) cannot be sustained without human inputs, therefore humans are a integral part of agro-ecosystems. Practices to increase agro-ecosystem stability is given below in Table 6.2.

Table 6.2. Practices to increase agro-ecosystem stability and function.

Stability Factor	Examples	Factors
	Disturbance (frequency and intensity)	
Chemical	Fertilizers and pesticides	Account for mineralization of organic amendments, non-target effects of pesticides
Biological	Introduction of exotic or weed species	
Physical	How often and what kind of tillage	Reduced, minimum or no-till practices
	Diversity	
Species	Genetic resources (crop) or competition for water and nutrients (weeds)	Intercropping of varieties
Structure or habitat	Variety of plant heights	Intercropping of species
Temporal	Variety of plants through time	Rotations
	Complexity	
Trophic groups	How many functions are represented	Practices that improve habitat for soil
Redundancy	How many populations perform each function	Organisms, such as organic matter amendments, reduced disturbances
Food web structure	How do all of these groups interact	Increased diversity of resources and niches
	Nutrient or energy flux	
Nutrient flux	How fast nutrients or energy move through the system	Low input, high organic matter
	Input: Output efficiencies	Eliminate over-applications

Resistance and resilience to disturbance are characteristics of a healthy soil. The ability to continue providing essential services in the face of disturbances, whether natural or human induced, is essential to maintain or improve soil quality over time. A soil is not considered "healthy" if it is managed for short term productivity at the expense of future degradation.

SOIL QUALITY AND SOIL HEALTH

Soil quality and soil health are terms describing similar concepts, but the latter appeals to farmers and crop consultants as part of a holistic approach to soil management. We regard soil health as the integration and optimisation of the physical, biological and chemical aspects of soils for improved productivity in an economic and sustainable manner.

Biological and chemical processes, such as root growth, organic matter turnover, macrofauna activity, bacterial and fungal proliferation, and cation charge density on the exchange complex influence pore size distribution, density and stability of soil structure. In turn, the physical structure of a soil also plays an integral role in controlling chemical and biological processes. A high proportion of stable aggregates in an agricultural soil is desirable, especially in fine- and medium-textured soils, as they sustain a range of pore sizes and promote aeration, water infiltration, and drainage, as well as better soil workability, seed-bed quality, and easier root penetration. Agricultural management practices such as tillage, crop rotation practices, cover crops and organic matter additions strongly influence the components of soil quality and thus crop performance.

In the past, an overemphasis on chemical soil management has resulted in the deterioration of the biological and physical processes in soils. Prior to the introduction of inorganic fertilizers, soil fertility management consisted mostly of adding organic nutrient sources to the soil. Nutrients held in the organic forms were mineralised to inorganic forms by the soil microbial population and then made available for crop uptake. With inorganic fertilizers, the nutrition of plants has been short-cut and the lack of organic additions in many non-livestock farming systems does not adequately sustain important biological processes and the associated physical benefits (e.g., aggregation). A new emphasis on soil health through the integration of the chemical, biological and physical processes therefore provides a useful and holistic framework to discuss soil management in the age of diverse cropping systems with tools such as organic and inorganic fertilizers, reduced tillage, cover cropping, new rotations and other soil/crop management practices.

Understanding of soil quality and the multiple interactions within its compartments requires research on soil properties and processes. This should include basic studies on the effects of different treatments and managements on soil, and should give insight into soil functions while serving as a means of assessment mean for selecting proper tools for soil quality monitoring. Monitoring on field scale would enable farmers to identify problems at an early stage, and help them to decide what measures are needed to eliminate or alleviate the factor that is impairing soil function and limiting productivity. The ultimate goal is the most profitable and environmentally sound long-term management system for farms and fields as a whole. Monitoring on regional and national scales, in turn, would offer information to policy makers and administration, and enable modelling.

SOIL QUALITY INDICATORS

Soil properties that seem to deliver the best information about the status of a soil are frequently called "indicators". The smallest set of indicators give information called the minimum data set. Indicators of soil quality may include soil characteristics, plant characteristics, and even economic calculations. Indicators may be qualitative (e.g. drainage is fast) or quantitative (infiltration= 2.5 in/hr).

Ideal indicators should:

- correlate well with ecosystem processes.
- integrate soil physical, chemical, and biological properties and processes.

- be accessible to many users.
- be sensitive to management and climate.
- be components of existing databases.

There are three main categories of soil indicators i.e, chemical, physical and biological. Typical soil tests only look at chemical indicators. Soil quality attempts to integrate all three types of indicators. These categories do not neatly align with the various soil functions. That is why integration is necessary. The relationship between indicator type and soil function is shown below:

Chemical :	Nutrient cycling, water relations, buffering
Physical :	Physical stability and support, water relations, habitat
Biological :	Biodiversity, nutrient cycling, filtering

Some indicators are descriptive and can be used in the field as part of a health card. Others must be measured using laboratory analyses. Some examples of indicators (and indicator groupings) that fall into the three broad categories of chemical, physical and biological are provided below:

Indicator Categories

Chemical indicators can give information about the equilibrium between soil solution (soil, water and nutrients) and exchange sites (clay particles, organic matter); plant health; the nutritional requirements of plant and soil animal communities, levels of soil contaminants and their availability for uptake by animals and plants. Indicators include:

- Organic matter measures, such as total organic carbon and total nitrogen.
- Soil fertility measures, such as soil test phosphorus, mineral nitrogen, exchangeable potassium, and micronutrients.
- Soil reaction (pH).
- Measures of toxic contaminants, such as arsenic or copper.
- Measures of acidity, salinity or sodicity, such as exchangeable aluminium, electrical conductivity, sodium adsorption

ratio, and exchangeable sodium percentage.

Physical indicators provide information about soil hydrologic characteristics such as water entry and retention which influences availability to plants. Some indicators are related to nutrient availability by their influence on rooting volume and aeration status. Other measures tell about erosion status. Indicators include measures of-

- Water flow into and through soil, such as porosity, water infiltration rate and water retention.
- Soil structure, such as porosity, bulk density, and soil depth.
- Soil aggregate size.
- Physical stability, such as aggregate stability, soil depth, or soil loss.

Biological indicators tell about the organisms that form the soil food web, which are responsible for decomposition of organic matter and nutrient cycling. Information about the numbers of organisms, both individuals and species, that perform similar jobs or niches, can indicate a soil's ability to function or bounce back after disturbance (resistance and resilience). The full consequences of extinction of organisms are unknown. Indicators include measures of-

- Active organic matter pools, such as microbial biomass carbon or respiration.
- Diversity indices for various soil organism populations, such as earthworms, nematodes, and microarthropods.
- Biological activity, such as enzyme activity, potentially mineralizable nitrogen or respiration (CO_2 production).

SOIL QUALITY ASSESSMENT

The need for a methodology for characterization of soil quality is being increasingly recognized. Earlier, soil quality was taken as a synonym for the capacity of soil to produce yield, that is, as a synonym for soil fertility. In recent years, soil quality is being seen to involve more than inorganic chemical soil tests and crop yield. Problems related to soil quality and function, other than nutrient deficiency, include poor water infiltration, crusting, erosion and poor biological activity, together resulting in poor nutrient cycling, reduced crop growth and soil degradation. Many of these problems are related to poor soil structure, which, in turn, is affected by biological activity.

Soil quality cannot be measured directly because it is a broad, integrative concept. Instead, we analyze a variety of proxy measurements that together provide clues about how the soil is functioning.

A healthy soil does perform five essential functions.

i. **Nutrient cycling.** Soil stores, moderates the release of, and cycles nutrients and other elements. During these biogeochemical processes, analogous to the water cycle, nutrients can be transformed into plant available forms, held in the soil, or even lost to air or water.

ii. **Water relations.** Soil can regulate the drainage, flow and storage of water and solutes, which includes nitrogen, phosphorus, pesticides, and other nutrients and compounds dissolved in the water. With proper functioning, soil partitions water for groundwater recharge and for use by plants and soil animals.

iii. **Biodiversity and habitat.** Soil supports the growth of a variety of plants, animals, and soil microorganisms, usually by providing a diverse physical, chemical, and biological habitat.

iv. **Filtering and buffering.** Soil acts as a filter to protect the quality of water, air, and other resources. Toxic compounds or excess nutrients can be degraded or otherwise made unavailable to plants and animals.

v. **Physical stability and support.** Soil has the ability to maintain its porous structure to allow passage of air and water, withstand erosive forces, and provide a medium for plant roots. Soils also provide anchoring support for human structures and protect archeological treasures.

Nutrient Cycling

Soil is the major "switching yard" for the global cycles of carbon, water, and nutrients. Carbon, nitrogen, phosphorus, and many other nutrients are stored, transformed, and cycled through soil.

Carbon dioxide and soil. The carbon cycle illustrates the role of soil in cycling nutrients through the environment. More carbon is stored in soil than in the atmosphere and above-ground biomass combined. Soil carbon is in the form of organic compounds originally created through photosynthesis in which plants convert atmospheric carbon dioxide (CO_2) into plant matter made of organic carbon compounds, such as carbohydrates, proteins, oils, and fibers. The organic compounds enter the soil system when plants and animals die and leave their residue in or on the soil. Immediately, soil organisms begin consuming the organic matter, extracting energy and nutrients and releasing water, heat, and CO_2 back to the atmosphere. Thus, if no new plant residue is added to the soil, soil organic matter will gradually disappear. If plant residue is added to the soil at a faster rate than soil organisms convert it to CO_2, carbon will gradually be removed from the atmosphere and stored (sequestered) in the soil. Cultivation aerates the soil, triggering increased biological activity, and therefore rapid decomposition, loss of soil organic matter, and release of CO_2 into the atmosphere. Most soil carbon losses occur in the first several years after cultivation begins. Farmers and other conservationists are interested in reversing that effect and increasing the amount of carbon stored in the soil. In general, reducing tillage can increase the extent of carbon sequestration and the amount of organic matter retained in the soil.

Soil organisms. Decomposition is a stepwise process involving virtually all soil organisms. Arthropods and earthworms chew the material and mix it with soil. A few fungi may break apart one complex compound into simpler components, then bacteria can attack the newly created compounds, and so on. Each organism gets energy or nutrients from the process. Decomposition liberates carbon and nutrients from the complex material making up life forms putting them back into biological circulation so they are available to plants and other organisms.

Decomposition also degrades compounds in soil that would be pollutants if they entered ground or surface water. The portion of plant and animal residue that is not broken down is transformed into the highly complex organic compounds called humic substances that can persist in soil for centuries and are important to soil structure and nutrient storage.

Thus, nutrient cycling process is an important phenomenon in the soil environment and need to be assessed by measuring the following indicators-

- Fertility Indicators including potentially mineralizable nitrogen (PMN), mineral nitrogen (N), soil test phosphorus (P), exchangeable potassium (K).
- Organic Matter Indicators including total organic carbon (TOC), microbial biomass carbon (MBC), particulate organic matter (POM), decomposition, carbon mineralization.
- Soil Reaction Indicators including soil solution pH (in water or calcium chloride).

Water Relations

When rain or irrigation water falls to earth, some of the water will infiltrate into the soil and some will flow over the surface. If the soil is loose, porous, and has a stable structure, a drop of water will be likely to infiltrate. If the soil has few openings and unstable structure a crust forms and seals the soil surface.

Plants are also important in determining the fate of water. Leaves intercept water, so some evaporates before it ever reaches the soil, and leaves and plant residue protect the soil so rain hits more gently. Roots and residue slow down the flow of water over land, so water has more time to soak in.

If the soil becomes saturated, some water will drain down to groundwater. The remainder will be held in the soil until it evaporates or is drawn into plant roots, eventually transpiring from leaves. At all these stages water is carrying sediment, organic matter, plant nutrients such as nitrogen and phosphorus, pesticides, and other dissolved or suspended compounds. Water flowing over the surface may carry sediment and nutrients into lakes. Water draining into groundwater may contain nitrate or pesticides.

Soil water relations can be assessed by measuring or observing the following indicators-

- *Physical Stability Indicators* including aggregate stability and erosion patterns.
- *Water Availability Indicators* including infiltration rate, available water capacity (AWC), water holding capacity (WHC), soil moisture, hydraulic conductivity, and ponding patterns.
- *Salinity and Sodicity Indicators* including electrical conductivity (EC), sodium absorption ratio (SAR), and exchangeable sodium percentage (ESP).

Biodiversity and Habitat

Each animal, plant, and microbe species requires a slightly different habitat. Thus, a wide variety of habitats are required to support the tremendous biodiversity on earth. At the microbial level, diversity is beneficial for several reasons. Many different organisms are required in the multi-step process of decomposition and nutrient cycling. A complex set of soil organisms can compete with disease-causing organisms, and prevent a problem-causing species from becoming dominant. Many types of organisms are involved in creating and maintaining the soil structure that is important to water dynamics in soil. Many antibiotics and other drugs and compounds used by humans come from soil organisms. Most soil organisms cannot grow outside of soil, so it is necessary to preserve healthy and diverse soil ecosystems if we want to preserve beneficial microorganisms. Estimated numbers of soil species include 30,000 bacteria; 1,500,000 fungi; 60,000 algae; 10,000 protozoa; 500,000 nematodes; and 3,000 earthworms.

The ability of soil to support plant and animal life can be assessed by measuring the following indicators-

- Biological Activity Indicators *including yield, yield stability, yield relative to inputs, enzymes microbial biomass, and fungal biomass.*

- Biological Diversity Indicators *including habitat diversity and diversity indices for organisms such as nematodes, microarthropods, bacteria.*

Filtering and Buffering

The minerals and microbes in soil are responsible for filtering, buffering, degrading, immobilizing, and detoxifying organic and inorganic materials, including industrial and municipal by-products and atmospheric deposits. Soil absorbs contaminants from both water and air. Some of these compounds are degraded by microorganisms in the soil. Others are held safely in place in the soil, preventing contamination of air and water. When the soil system is overloaded, such as with the excess application of fertilizer or manure, or when the soil is unstable, some contaminants will be released back to the air and water through erosion or leaching. Wetlands function as nature's filters. Many human-made wetlands are created for this purpose.

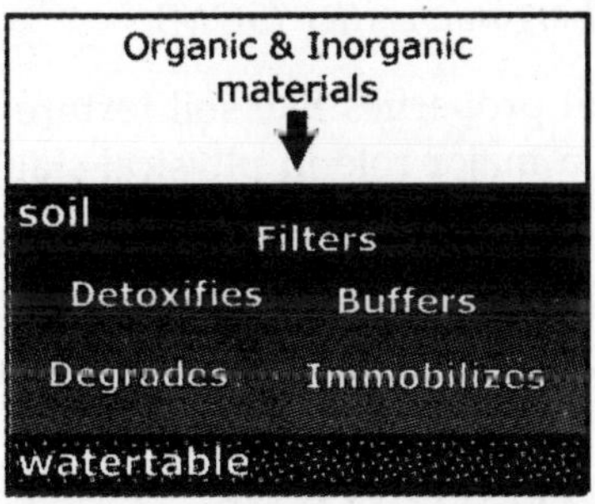

The filtering function of soil can be assessed by measuring or observing the following indicators-

- Toxicity Indicators
- Organic Matter Indicators
- Soil Reaction Indicators
- Salinity and Sodicity Indicators
- Biological Diversity Indicators

Physical Stability and Support

Soil support is necessary to anchor plants and buildings. Both flexible (it can be dug) and stable (it can withstand wind and water erosion). Soil also provides valuable long-term storage options including protecting archeological treasures and landfilling human garbage. The need for structural support can conflict with other soil uses. For example, soil compaction may be desirable under roads and houses, but can be devastating for the plants growing nearby.

The stability and support function of soil can be assessed by measuring the following indicators-

- Physical Stability Indicators *including aggregate stability and erosion patterns.*
- Soil Structure Indicators *including bulk density (BD), porosity, or penetration resistance.*
- Aggregate Size and Stability Indicators *including the mean weight diameter of water stable aggregates (MWD) or soil slaking indices.*
- Organic Matter Indicators *including total organic carbon (TOC) or soil organic matter (SOM).*

Also, inherent soil properties like soil texture and particle size distribution play a major role in physical stability.

Assessments based on indicators requiring more specialized equipment or more precise measurement, such as microbial biomass carbon or soil-test phosphorus. These include the Soil management assessment framework.

SOIL QUALITY MANAGEMENT

To improve the soil productivity, water quality, and other soil benefits, starts with following six components of soil quality management. To assess the effect of existing land management practices, identification of practices which fall into each of these categories is essential. Choosing specific practices within each category depends on field situation because different kinds of soil respond differently to the same practice.

Enhance Organic Matter

Whether soil is naturally high or low in organic matter, adding new organic matter every year is the most important way to improve and maintain soil quality. Regular additions of organic matter improve soil structure, enhance water and nutrient holding capacity, protect soil from erosion and compaction, and support a healthy community of soil organisms. Practices that increase organic matter include- leaving crop residues in the field, choosing crop rotations that include high residue plants, using optimal nutrient and water management practices to grow healthy plants with large amounts of roots and residue, growing cover crops, applying manure or compost, using low or no tillage systems, and mulching.

Avoid Excessive Tillage

Reducing tillage minimizes the loss of organic matter and protects the soil surface with plant residue. Tillage is used to loosen surface soil, prepare the seedbed, and control weeds and pests. But tillage can also break up soil structure, speed the decomposition and loss of organic matter, increase the threat of erosion, destroy the habitat of helpful organisms, and cause compaction. New equipment allows crop production with minimal disturbance of the soil.

Manage Pests and Nutrients Efficiently

An important function of soil is to buffer and detoxify chemicals, but soil's capacity for detoxification is limited. Pesticides and chemical fertilizers have valuable benefits, but they also can harm non-target organisms and pollute water and air if they are mismanaged. Nutrients from organic sources also can pollute when misapplied or over-applied. Efficient pest and nutrient management means testing and monitoring soil and pests; applying only the necessary chemicals, at the right time and place to get the job done; and taking advantage of non-chemical approaches to pest and nutrient management such as crop rotations, cover crops, and manure management.

Prevent Soil Compaction

Compaction reduces the amount of air, water, and space available to roots and soil organisms. Compaction is caused by repeated traffic, heavy traffic, or traveling on wet soil. Deep compaction by heavy equipment is difficult or impossible to remedy, so prevention is essential.

Keep the Ground Covered

Bare soil is susceptible to wind and water erosion, and to drying and crusting. Ground cover protects soil, provides habitats for larger soil organisms, such as insects and earthworms, and can improve water availability. Ground can be covered by leaving crop residue on the surface or by planting cover crops. In addition to ground cover, living cover crops provide additional organic matter, and continuous cover and food for soil organisms. Ground cover must be managed to prevent problems with delayed soil warming in spring, diseases, and excessive build-up of phosphorus at the surface.

Diversify Cropping Systems

Diversity is beneficial for several reasons. Each plant contributes a unique root structure and type of residue to the soil. A diversity of soil organisms can help control pest populations, and a diversity of cultural practices can reduce weed and disease pressures. Diversity across the landscape can be increased by using buffer strips, small fields, or contour strip cropping. Diversity over time can be increased by using long crop rotations. Changing vegetation across the landscape or over time not only increases plant diversity, but also the types of insects, microorganisms, and wildlife that live on your farm.

SUMMARY

The welfare of the present and future population depends on soil quality and resilience under an intensive agriculture use. Agricultural land management in the past has stressed the soil system, often withdrawing more than what was returned. In particular continuous cultivation and intensive cropping have resulted in loss of soil by erosion, decrease in soil organic matter and a concomitant release of carbon as CO_2 into atmosphere.

A clear merit of the soil quality concept and assessment is the integration of important but often separately considered aspects of soil. In the agricultural context, the integrative approach is highly useful in producing the knowledge needed by the various stakeholders of arable land management. The assessment of soil quality is invaluable in determining the sustainability of soil and land management systems and in evaluating the long-term effectiveness of the systems. Internationally, the implementation of the soil quality approach has improved the educational competence of the scientific community and increased the transparency of environment related science to the surrounding society.

7

Environmental Concerns Associated with Fertilizer Use

Use of fertilizer materials has generated numerous environmental concerns in recent years. Concerns can be categorized by their effect on water quality, air quality, and human and animal health. In each case, constituents of primary interest are nitrogen and phosphorus, although others need to be considered depending on the fertilizer source. There are many available fertilizer sources including commercial fertilizers, biosolids and animal waste. Environmental concerns become a potential hazard with the misuse of these materials. Misuse generally arises when fertilizer application rates exceed agronomic requirements. It is emphasized here that application of fertilizer materials is not environmentally unsound but excessive application of any of them can lead to potential hazards. Therefore, an agriculture systems manager should be aware of potential problems. By knowing the potential problems, fertilizer inputs can be properly managed to maximize production and minimize negative environmental impacts.

FERTILIZER MATERIALS

Fertilizer materials are either straight materials (those containing only one nutrient) or mixtures (those containing two or more nutrients). Mixed dry materials are available in two forms: i) chemical compounds in which two of the major fertilizer elements are combined together in the granule, and ii) bulk blends in which straight materials and/or chemical compounds are physically blended to make various grades. The major fertilizer materials available in for farming communities are listed in Table 7.1.

Table 7.1. Major fertilizer sources of nitrogen, phosphorus and potassium.

Source	Nutrient Composition (%)						
	N	P_2O_5	K_2O	CaO	MgO	S	Cl
Nitrogen							
Ammonium sulphate	21	-	-	-	-	24	-
Anhydrous ammonia	82	-	-	-	-	-	-
Ammonium nitrate	33-34	-	-	-	-	-	-
Calcium nitrate	15	-	-	34	-	-	-
Urea	45-46	-	-	-	-	-	-
Urea-ammonium nitrate (solution)	28-32	-	-	-	-	-	-
Phosphorus							
Monoammonium phosphate (MAP)	11	48-55	-	2	0.5	1-3	-
Diammonium phosphate	18-21	46-54	-	-	-	-	-
Ammonium poly-phosphate (solution)	10-11	34-37	-	-	-	-	-
Urea-phosphate	17	43-44	-	-	-	-	-
Ordinary super-phosphate	-	16-23	-	18-21	-	11-12	-
Triple super-phosphate	-	44-53	-	12-14	-	0-1	-
Rock phosphate	-	25-40	-	33-36	-	-	-
Potassium							
Potassium chloride	-	-	60-62	-	-	-	47
Potassium sulphate	-	-	50-52	-	-	17	-

Nitrogen Fertilizers

Anhydrous Ammonia, NH_3, 82% N. Nitrogen was one of the first nutrients to be produced in a liquid form (liquid under pressure). Nitrogen is taken from the air and reacted with a hydrogen source in the presence of a catalyst to produce anhydrous ammonia. Virtually, all nitrogen manufacturing facilities use natural gas as a source of hydrogen. Approximately 33,000 cubic feet of natural gas are required to produce a ton of ammonia. Under pressure, anhydrous ammonia becomes a liquid that returns to a gas when released from the storage container. To prevent excessive loss of N, it must be injected into the soil and sealed until ammonium (NH_4^+) is formed. Anhydrous ammonia is a hazardous material and care must be taken in handling to avoid exposing human, animal or plant life to direct contact with liquid or gaseous forms. In nitrogen producing plants, anhydrous ammonia is the basic material used to produce other kinds of nitrogen fertilizers.

Urea ammonium-nitrate, 28-32% N. A common liquid N fertilizer is made from soluble urea and ammonium nitrate mixed in equal parts with water to form non-pressure N solution containing 28 to 32 percent nitrogen. Ammonium nitrate or urea solution, alone, can only be handled satisfactorily in the field, in approximately 20% N concentrations. Like any salt solution, nitrogen solutions is salt out. Salting out is simply the precipitation of the dissolved salts when the temperature drops to a certain degree. The salting out is determined by the amount and kind of salts in solution. As a general guide, 28% non-pressure solution salts out at about 0°F and 32% salt out at about 32°F, although this can vary between the materials produced by different manufacturers. Nitrogen solutions that do not contain free ammonia can be applied to the soil surface without loss of N, although incorporation is recommended where ammonia volatilization loss from urea may be a problem. Ammonia free N solutions can also be applied in sprinkler irrigation systems with good success. Non-pressure N solutions are probably the most versatile of all N materials for application to a broad range of crops with a wide variety of application equipment.

Ammonium Nitrate, NH_4NO_3, 33.5-34% N. Ammonium nitrate is made by reacting anhydrous ammonia and nitric acid. Half of the total nitrogen in the material is in the nitrate form and half is in the ammoniacal form. Most ammonium nitrate is prilled and coated.

Urea, $(NH_2)_2CO$, 45-46% N. Urea is formed by reacting ammonia and carbon dioxide. All of the nitrogen in urea is in the ammoniacal form. Urea is produced in both prilled and granular forms. It is classed as an organic compound since it contains carbon.

Ammonium Sulphate, $(NH_4)_2SO_4$, 20.5-21% N. Ammonium sulphate is formed by reacting ammonia with sulphuric acid. All of the material's nitrogen is in the ammoniacal form. Ammonium sulphate is an effective source of sulphur, since it contains 24 percent S. It is produced in both crystalline and granular forms.

Phosphorus Fertilizers

Diammonium Phosphate, DAP, $(NH_4)_2HPO_4$, 18% N, 46% P_2O_5. This popular N-P material is produced by reacting ammonia and phosphoric acid. All of the nitrogen is in the ammoniacal form and the P is highly water-soluble. It is produced in the granular form.

Monoammonium Phosphate, $NH_4H_2PO_4$, 11-12% N, 48-60% P_2O_5. This material is produced by reacting ammonia and phosphoric acid. All of the N is in the ammoniacal form and the P is highly water-soluble. Most MAP is produced in the granular form.

Phosphoric Acid and Superphosphoric Acid, 54-85% P_2O_5. Phosphate rock deposits are the basic source of all phosphate materials. The principal world reserves are located in North Africa, North America and the former Soviet Union. The primary intermediate step in the production of phosphorus fertilizers is phosphoric acid. In some areas, phosphoric acid is applied to the soil as a form of fertilizer; however, the handling problems associated with this acid has limited its use. In fluid fertilizer production, two types of acid are commonly used; ortho phosphoric (phosphoric acid) containing about 54% phosphorus (P_2O_5) and superphosphoric (polyphosphoric acid) containing up to 85% phosphorus (P_2O_5). Being more concentrated, it is possible to produce a higher analysis P fertilizer from superphosphoric acid. When ortho phosphoric acid is reacted with ammonia, the acid can be neutralized to a pH of about 6.5 to produce a nitrogen phosphorous solution. This was the basic phosphorous material used in mixed liquid fertilizers for several years. The development of superphosphoric production procedures makes it possible to produce the higher analysis nitrogen phosphorous solutions (10-34-0), currently used as the basic phosphorous source in liquid and suspension grades of liquid fertilizer.

Ammonium Polyphosphate Solutions, APP, 10% N, 34% P_2O_5. Ammonium polyphosphate solutions can contain up to 70 percent of the total P_2O_5 as a poly-P form. The remaining P_2O_5 is as an orthophosphate. All phosphate fertilizers contain some orthophosphate with many being 100% in the ortho form. In fluids, it is generally accepted that high poly content, above 55 percent, improves storage quality and the opportunity to carry low cost sources of micronutrient metals in liquid grades.

Ordinary Superphosphate, 20% P_2O_5. Ordinary superphosphate is made by treating finely ground phosphate rock with sulphuric acid. The P_2O_5 content of this source ranges between 18 and 22 percent. This source has between 11 and 12 percent sulphur as calcium sulphate and is sold as granular form.

Concentrated Superphosphate, 46% P_2O_5. This source is produced by treating ground rock phosphate with phosphoric acid. The product will vary from 42-46 percent P_2O_5 with the most common analysis 46% P_2O_5.

Potassium Fertilizers

Potassium (K) is found throughout the world in both soluble and insoluble forms. The soluble forms are the principal form used in fertilizers. Potassium chloride is by far the most important source of fertilizer K.

Potassium Chloride (Muriate of Potash), KCl, 60% K_2O. This is the K salt of hydrochloric (muriatic) acid. Most potash deposits are in this form. It is the most popular potash material used in fertilizers. Muriate of potash is a crystalline material. It is available in various particle sizes which are chosen to coincide with other materials for bulk blending. Some muriate of potash contains iron coatings, giving it a reddish colour. Most muriate of potash is white or translucent. Colour or particle size does not affect potassium availability for plant growth, since it is a water soluble compound. In addition, potassium chloride is the major source of potash for liquid fertilizers.

Potassium Sulphate, K_2SO_4, 50% K_2O. Like muriate of potash, potassium sulphate occurs naturally in limited deposits. It contains 17 percent sulphur and is widely used in areas where both potassium and sulphur are needed. Potassium sulphate has a lower solubility than KCl and is primarily used in suspensions to produce chloride free potassium and sulphur.

Secondary Elements

Calcium (Ca). Common sources of supplemental Ca are lime and gypsum.

Calcium Carbonate (Lime)	20-40% Ca
Calcium Sulphate (Gypsum)	23% Ca, (18.6% Sulphur)
Normal Superphosphate	22% Ca, (20% P_2O_5, 12% S)

Magnesium (Mg). The most common sources of magnesium are magnesium sulphate and dolomitic lime.

Magnesium Oxide	52% Mg
Magnesium Sulphate	16% Mg
Potassium - Magnesium Sulphate	11% Mg, (22% K_2O, 22% S)
Dolomitic Limestone (varies)	12% Mg

Sulphur (S). Sulphur is most available when supplied in the highly water soluble sulphate form. Elemental sulphur can be used, but requires biological oxidation over time to convert the elemental form to available sulphate.

Calcium Sulphate (Gypsum)	17% S(22% Ca)
Potassium Sulphate	17% S
Sulphate of Potash, Magnesia	22% S
Ammonium Sulphate	24% S

Normal Superphosphate	12% S
Ammonium Thiosulphate	26% S

Micronutrients

Zinc (Zn), Iron (Fe), Copper (Cu), and Manganese (Mn). The micronutrient elements can be discussed as a group, since their sources are somewhat similar. Industry separates the compounds into two general categories; inorganic and organic. Inorganic include sulphates, oxides, carbonates and chlorides. The term organic applies primarily to chelated products and some sequestered materials. Most chelates, and particularly liquid products, can be mixed with liquid without difficulty.

Zinc

Zinc Sulphate	25-36% Zn
Zinc Oxide	50-80% Zn
Zinc Chloride	48% Zn
Zinc Chelate	9-14.5% Zn

Iron

Ferrous Sulphate	20.1% Fe
Ferric Sulphate	19.9% Fe
Ferrous Ammonium Sulphate	14.2% Fe
Ferric Chloride	34.4% Fe
Iron Chelate	10% Fe

Copper

Copper Sulphate	25% Cu

Manganese

Manganese Sulphate	23-28% Mn

Boron (B). A sodium borate (solubor) containing about 20% B is the source of B most commonly used in liquids. Boric acid and other soluble forms containing between 14 to 20% B are also suitable for liquid mixes.

Borax	11.3% B

Molybdenum (Mo). Ammonium molybdate is satisfactory for liquids. Sodium molybdate is used, although it is less soluble than ammonium molybdate.

Sodium Molybdate	39.7% Mo
Ammonium Molybdate	54.3% Mo

ENVIRONMENTAL CONCERN

Nitrogen

Environmental concerns with N focus on water quality but also include air quality and human and animal health. Water quality issues include N concentrations in surface water and groundwater. Concerns for surface waters are related to N entering streams, ponds, and lakes where elevated levels will stimulate algae growth resulting in algae blooms. Upon the death of the algae, microbial activity increases resulting in a decrease in available oxygen for biological functions, a condition referred to as *eutrophication.* Eutrophication has a detrimental effect on most aquatic species. It occurs when there are adequate sources of nutrients, but the system is limited by the available oxygen, resulting in the death of many aquatic species including fish and invertebrates.

The most common pathway for land applied N to reach surface waters is by runoff waters. These waters will often contain soluble materials and soil sediments. Therefore, even N applied at agronomic rates and incorporated into the soil is susceptible to moving into surface waters by runoff when carried by soil particles. Nitrate-N is a soluble N form and ammonium-N can be attached to the soil particles as they are carried into the stream or impoundment.

To minimize N problems associated with runoff from fields into surface waters several steps can be taken. One of the most effective is to maintain plant residue on the soil surface which will enhance water infiltration and reduce the amount of soil sediments moved from the field into surface water. Another effective practice is to leave a buffer strip of vegetation between the field and the surface water, which can act as a trap for many of the soil sediments. By catching sediments in the buffer strip the amount of N reaching the surface water is reduced.

Although, eutrophication of surface waters is important, much of the regulation focuses on the use of N in areas where a subsurface aquifer is within 3 metre of the soil surface. Nitrogen in the NO_3^- form is very susceptible to leaching through the soil profile, therefore, these sites possess a real possibility for elevated levels of NO_3^- to enter the aquifer when N application rates are in excess of agronomic rates. Concerns with nitrate reaching an aquifer are generally related to animal and human health rather than an imbalance in environmental nutrient requirements.

Methemoglobinemia (blue-baby syndrome) can result from the ingestion of nitrate in water or nitrate-rich food products. Ingested nitrate can then be reduced to nitrite in the upper gastro-intestinal tract, and once incorporated in the blood system can form methemoglobin. Methemoglobin, unlike hemoglobin, cannot function as an oxygen carrier ultimately resulting in anoxia or suffocation, if high amounts are present. Infants younger than 3 months are highly susceptible to gastric bacterial nitrate reduction because they have very little gastric acid production and low activity of the enzyme that reduces methemoglobin back to hemoglobin.

N-nitrosamines are potent carcinogens in animals. These compounds can be synthesized from amines and nitrous acid under certain conditions. When nitrate is reduced to nitrite it can give rise to the formation of N-nitrosamine compounds that are an important class of chemical carcinogens for humans. However, nitrosamines occur in very few foods and at very low levels because of their chemical instability. It is important to note that the presence of nitrosamines in food products is generally not associated with nitrates from N fertilizers, but rather the use of nitrite as a curing agent in meats, poultry, and fish. Potassium nitrate has also been used as a food preservative. Other studies have shown an association between nitrate in drinking water and the incidence of gastric carcinoma in adults continuously exposed to high nitrate.

Agronomic solutions have been available for years to deal with fertilizer NO_3-N pollution of surface and subsurface water supplies. Nitrogen fertilizer recommendations based on removal and use efficiency have been shown to be both environmentally sound and economical. Recent research on soil fertility project has demonstrated limited potential for NO_3-N leaching when the recommended N fertilization rates are employed in continuous winter crops. This work has also shown that N rates needed for maximum grain yield can be exceeded by small amounts without increasing soil profile NO_3-N accumulation.

The use of N in agriculture has been identified as a contributor to water pollution. However, it also has been found that this contribution to ground water contamination occurs when N is managed improperly. Under continuous production of crops, applied N at the recommended rate (using soil testing and realistic yield goals) will not result in increased NO_3-N contamination of groundwater. By working at a sub-field scale, excessive N application can be reduced, thus reducing the risk of NO_3-N leaching to groundwater.

A final concern related to the use of N fertilizers in some regions is air quality. This is primarily related to the application of animal manures and biosolids and resulting odour associated with them. There could be a push to regulate land application of animal manures and biosolids based on the NH_3 associated with them. Some believe a potential exists for degradation of air quality and detrimental effects to human health. This could be extended to the application of ammonium and ammonia containing commercial fertilizers as well. To minimize concerns associated with air quality, it is recommended that ammonia-containing fertilizers be incorporated upon application. There are agronomic and financial reasons for doing this as well as those associated with air quality. By incorporating these fertilizer sources, the amount of N lost from the soil system is reduced, thus, saving on the quantity of fertilizer purchases or allowing more land area to be fertilized with animal manure or biosolids.

Phosphorus

Environmental concerns with phosphorus focus on water quality, particularly surface water quality. Phosphorus in the soil is an immobile plant nutrient and is tightly adsorbed to soil particles significantly reducing leaching movement through the soil profile. Therefore, if phosphorus is to reach surface water, it must be transported by the sediment load in runoff waters. If phosphorus does reach a stream or other body of surface water, it can lead to the accelerated eutrophication of the recipient water body. As previously discussed, eutrophication is the condition where a body of water has an enriched nutrient load (phosphorus) but is limited by the available biological oxygen in the water. Algal species that proliferate in high phosphorus water include Anabaena, Ankstrodemus and Euglena. As these organisms die and are decomposed by other organisms, the available biological oxygen is significantly reduced causing adverse effects on other species of aquatic life. To reduce these adverse effects, proper application is needed.

Due to their immobility in the soil, nearly all commercial P fertilizers are incorporated after broadcast application or banded below the seed. To reach surface water, this source of P is transported in the sediment as well as in dissolved form in the runoff. Therefore, reducing runoff and erosion will reduce environmental concerns related to P. The most effective way to do this is to follow good soil conservation practices. These include increasing water infiltration, reducing runoff by maintaining surface residues and using buffer strips at the edge of the field. These good conservation practices allow you to maintain your fertilizers, reduce soil loss and increase water stored in the soil profile.

Land application of animal manures, particularly poultry litter (high in P), and some biosolids are done by broadcasting the material on the soil surface. In many cases, these fertilizer materials are applied to forage crops which eliminates their incorporation. When left on the surface in this manner, they may be subject to loss from the field in the runoff. Application rate on crop P needs instead of N needs will slow down P build up in the soil. To reduce P loss is to use a buffer strip at the edge of the field to reduce the amount of sediment and manure leaving the field.

Other Contaminants

With the decrease in suitable landfill sites for human waste and the increase in confined animal feeding operations, there has been a tremendous increase in the interest of land application of these materials. These materials should be viewed as a valuable nutrient source and not a waste material. They contain many plant nutrients in addition to N and P, so farmers, who have them, should use them to their maximum benefit. To date, no other constituents in these fertilizer sources have proven to be of major environmental concern when proper guidelines are followed. Each source has a different make-up due to ration formulation of materials in the municipal waste stream. Constituents which may need to be considered are copper (Cu) in animal waste and heavy metals in biosolids. Heavy metal concentrations of biosolids must be monitored with materials above threshold levels needing to be land filled.

SUMMARY

Environmental concerns due to the application of fertilizers can be drastically reduced by proper management of these resources. Regardless of fertilizer form, if the quantity applied is greater than what is required for the crop then the potential exists for negative environmental impacts.

To minimize negative environmental impacts, there are a few simple practices one can use:

- add only the amount of fertilizer needed to meet plant requirements,
- use buffer strips and do not apply fertilizers too close to bodies of water, and
- use good soil conservation practices which minimize soil erosion and maximize water infiltration.

A combination of these good management practices will greatly reduce the potential for adverse environmental impacts.

8

Heavy Metals Contamination in Soils

Heavy metals are defined as metals having density greater than 5 g/cm^3. This classification includes transition metals and higher atomic weight metals of groups III to V of the periodic table. These include aluminum, arsenic, beryllium, bismuth, cadmium, chromium, cobalt, copper, lead, mercury, nickel, selenium, strontium, titanium and zinc. These metals and their compounds may be discharged from industries, farmlands, municipal urban runoffs, etc, into surface waters to cause pollution. The pollution of the ecosystem by toxic metals poses serious concern in two ways. Firstly, metals are non-degradable and are persistent in the ecosystem. Secondly, physical, chemical and biological processes may combine under certain circumstances to concentrate metals rather than dilute them.

Heavy metals such as lead, arsenic, cadmium, copper, zinc, nickel, and mercury are continuously being added to our soils through various agricultural activities such as agrochemicals usage and long-term application of urban sewage sludge in agricultural soils, industrial activities such as waste disposal, waste incineration and vehicle exhausts, as well as from anthropogenic sources. All these sources cause accumulation of metals and metalloids in our agricultural soils and pose threat to food safety issues and potential health risks due to soil to plant transfer of metals. Co-existence and persistence of heavy metals in soils as multiple contaminants and human exposure to them through ingestion of heavy metal contaminated food or uptake of contaminated drinking water can lead to their accumulation in humans, plants and animals. They can also cause a considerable detrimental effect on soil ecosystems, environment and human health due to their mobilities and solubilities which determine their speciation. In some cases, the soil may be contaminated to such an extent that it may be classified as a hazardous waste. Soil contamination with heavy metal mixtures is receiving increasing attention from the public as well as governmental bodies, particularly in developing countries. The remediation of such soils is important because these usually cover large areas that are rendered unsuitable for agricultural and other human use.

SOURCES OF HEAVY METALS

There are many sources of the metallic contaminants that can accumulate in soils. The burning of fossil fuels, smelting and other processing techniques release into the atmosphere tons of these elements, which can be carried for miles and later deposited on the vegetation and soil. Lead, nickel and boron are gasoline additives that are released into the atmosphere and carried to the soil through rain and snow. Borax is used in detergents, fertilizers and forest fire retardants, all of which commonly reach the soil. Super phosphate and limestone, two widely used fertilizer and amendments, usually contain small quantities of cadmium and copper. Arsenic was for many years used as an insecticide on cotton, tobacco, fruit crops, lawns and as a vine killer.

In municipal sewage, the metallic contents are often absorbed on the sewage solids or sewage sludge. When the sludge is disposed off to farm land, the metallic contents are taken up by plants in some amounts. These may have unpleasant effect on the fruits they produce as they may be rendered unsuitable for human consumption. In some cases, they may have adverse effects on plants produced on the farm lands. Such plants can be readily consumed by grazing animals. The grazing animals as well as fruits of plants serve as food for man. In effect, the heavy metals are passed on to man through the food chain and the cumulative effects of these metals, most of which are toxic, are adverse generally.

Actually, some of the toxic metals are being released to the environment in increasing amounts, while others, most notably, lead, because of change in gasoline formulation, are decreasing. All are daily ingested by humans either through air, food, water and soil. Irrespective of their sources, toxic elements can and do reach the soil where they became part of the food chain. Unfortunately, once the elements become part of this cycle, they may accumulate in animal and human body tissue to toxic levels. This situation is especially critical for fish and other wildlife and for humans at the top of the food pyramid.

PHYTOAVAILABILITY OF HEAVY METALS IN SOIL

Heavy metal interactions in soils vary considerably with the nature of soil types. The phytoavailability of metals is determined by the nature of the metal species, their interaction with soil colloids, the soil characteristics and duration of contact with the surface binding these metals. Soil characteristics (eg. soil pH, clay, organic matter content and type, and moisture content) also determine availability to plants by controlling the speciation of the elements, temporary binding by particle surfaces (adsorption-desorption processes), precipitation reactions and availability in soil solution. Both the concentration of heavy metals and their speciation vary significantly with the composition of soil solution and the amount of moisture present in the soils. Plants growing in different soils with the same total metal concentration may vary in their phytotoxic response due to the differences between soils in their sorptive capacity. The different amount of metals retained by different soil types may be attributed to the different clay content and mineral composition of the soils, their pH, organic matter content and soil solution composition which is the medium of reaction in soils. Fertilizer like rock phosphate is a highly potential source of lead and cadmium. In soils with coarser textures and low pH (acidic reaction), heavy metals, applied through such fertilizers, are available more than in those containing large amount of clay and with high pH (alkaline reactions).

Replenishment Capacity of Soil

Heavy metals are associated with soil solids in four major ways:

i) A very small proportion is held in adsorbed or exchangeable forms which are available for plant uptake.

ii) The elements are bound by the soil organic matter and by the organic materials in the sludge. A high proportion of copper is commonly found in this form, while lead is not so highly attracted. Organically bound elements are not readily available to plants, but can be released over a period of time.

iii) The association of heavy metals in soil is with carbonates and with oxides of iron and manganese. These forms are less available to plants than either the exchangeable or organically bound forms, especially if the soils are not allowed to become too acidic.

iv) The fourth association is commonly known as the residual form, which consists of sulphides and very insoluble compounds that are less available to plants than any of the other forms.

Soil and plants do not readily adsorb/absorb applied heavy metals but, the heavy metals are also not easily leached from the soil. Thus, the immobility of the metals means that they accumulate in soils if repeated applications are made.

The replenishment factor is controlled by clay content, mineral and organic matter composition, partition coefficient and the chemical saturation index. In soils with high reaction capacity (high clay content soils) but low saturation index (i.e. binding sites are unsaturated with respect to metals), the phytoavailable fraction is low. Metal uptake is usually much higher in low sorbing sandy soils compared to high sorbing clayey soils, although both soils had similar total metal content. However, caution must be exercised with such conclusions as metal phytoavailability may vary with soil pH where decreasing pH decreases the binding capacity of metals and also with plant type.

Plant Species and Metal Phytoavailability

The ability of plants to bioaccumulate metals and possibly other contaminants varies with both the nature of plant species and the nature of metal contaminants. Laboratory studies consistently demonstrate that the capacity of plants to bioaccumulate metals varies extensively with the nature of metals as well as with plant types.

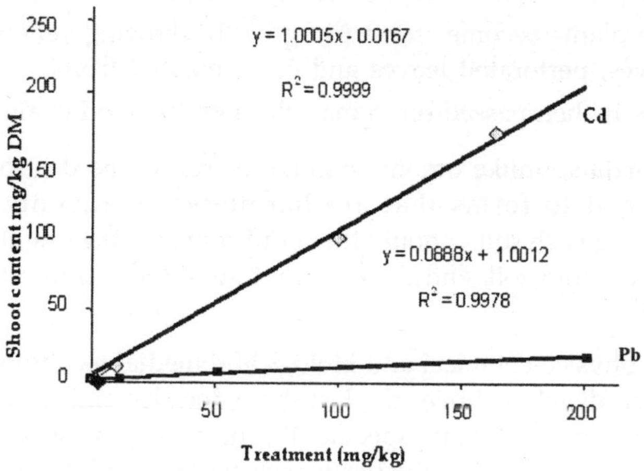

Figure 8.1. The capacity of plants (pisum sativum) to bioaccumulate metals.

Figure 8.1 illustrates the marked differences in the uptake of Cd and Pb by peas (Pisum sativum). These differences in metal uptake may be attributed to both the markedly different binding capacity of soils for these metals and also to plant-root-metal interactions, which vary with metal types. Whilst variances in metal uptake capacity of plants may have limited significance to metal phytotoxicity guidelines, the marked differences in the capacity of plants to bioaccumulate metals will have major implications to guidelines which are neither soil type specific nor plant type specific.

REMEDIATION TECHNOLOGY

Accumulation of heavy metals in soil has adverse effects on the plants:

- They cause nutrient depletion in the soil.
- The availability of nutrients in the soil for plant uptake is reduced.
- Those that mimic some nutrients tend to take up their place during plant uptake.
- The solubility of these nutrients becomes affected, as some of the contaminants are soluble.
- The plants suffer from deficiency of these nutrients.
- The plants become stunted in growth, showing yellowish leaves, perforated leaves and contaminated fruits.
- This is then passed on to man through the food chain.

Heavy metals, unlike organic pollutants, cannot be destroyed or changed to forms that are harmless. Treatments for remediating such soils, should therefore, aim at extracting these substances from soil, and then concentrating them before final disposal.

Various physico-chemical and biological remedial technologies have been developed over the last three decades and selection of each technology is site specific. Biochemical processes such as bioleaching involving *Thiobacillus spp.* bacteria and *Aspergillus niger* fungus, biosorption of low concentrations of metals in water by algal or bacterial cells, bio-oxidation or bio-reduction of metal contaminants by *Bacillus subtilis* and sulphate reducing bacteria, and biomethylation of metals such as arsenic, cadmium, mercury or lead, have shown some promises and could be used for soil sediment treatments. Several technologies exist for *in situ* chemically enhanced soil flushing by extracting solutions such as organic and inorganic acids, and complexation agents have also been proposed for remediation. All these methods in many cases are expensive, labour intensive, and result in extensive changes to the physical, chemical, and biological characteristics of the treated soil.

Phytoremediation

The health hazards associated with soil contamination with heavy elements having toxic effects together with high cost of removal and replacement of polluted soil have prompted to develop alternative and cheaper technologies to recover the degraded land. Current research in this area now includes plants to remediate polluted soils and to facilitate improvement of soil structure, the innovative technique being known as phytoremediation. Use of plants that have constitutive and adaptive mechanisms for tolerating or accumulating high metal contents in their rhizosphere and tissues, is the emerging in situ remediation technology employed in the clean up of soils, sediments and water that have been polluted by organics and heavy metals. It aims to use metal accumulating plants to remove, transfer or stabilize these contaminants from soil, sediments and water.

Phytoremediation can be categorized under five major subgroups:

(i) **Phyoextraction** – removal and concentration of metals into harvestable plant parts.

(ii) **Phytodegradation** – degradation of contaminants by plants and their associated microbes.

(iii) **Rhizofilteration** – absorption of metals by plant roots from contaminated waters.

(iv) **Phytostabilization** – immobilization and reduction in the mobility and bioavailability of contaminants by plant roots and their associated microbes.

(v) **Phytovolatilization** – volatilization of contaminants by plants from the soil into the atmosphere.

Phytoremediation, however, is a relatively slow process. It may take some years to reduce metal contents in soil to a safe and acceptable level due to small size and slow growth of most identified metal hyperaccumulator plants. To make phytoremediation a viable technology, there is need to either find fast growing and metal tolerant and or hyperaccumulating plants with extensive root system, or, genetically engineered common plants with as yet unidentified hyperaccumulation genes. Important and widely reported hyperaccumulators used for metal remediation is given in Table 8.1.

Table 8.1. Plant species for hyperaccumulation of heavy metals.

Heavy metals	Plant species	Maximum reported concentration (mg/kg)
Cadmium	*Thlaspi caerulescenes*	500
Copper	*Ipomoea alpine*	12300
Cobalt	*Haumaniuastrum robertii*	10200
Lead	*Thlaspi rotundifolium*	8200
	Brassica juncea	
	Zea mays	
Nickel	*Alyssum lesbiacum*	47500
	Sebertia acuminate	
Zinc	*Thlaspi caerulescenes*	51600
	Brassica juncea	
	B. oleracea	
	B. campestris	
Selenium	*Brassica juncea*	900
	B. napus	
Chromium	*Brassica juncea*	1400
	Helianthus annus	

Many fast growing and high biomass producing plants such as vetiver grass and hemp may not be defined as metal hyperaccumulators, but are metal tolerant allowing them to grow in soil with high metal concentrations. The possibilities of using such plant species which are easily growing in different climates, and using their biomass in non-food industries, can make them ideal plants for phytoremediation purposes. Phytoremediation must be considered as a long-term strategy. Various means have been tested in the last decade or so to find ways of enhancing the process.

Plant uptake of metals is frequently restricted by limitations of contaminant bioavailability and in order to enhance metal uptake, soil amendments with metal chelating agents such as EDTA, HEDTA, DTPA, EGTA, NTA, citrate and hydroxylamine to make metals bioavailable and absorbed by plant roots have shown promises. The type of chelate and its time of application are important considerations. It has also been suggested that if the plant biomass can be increased, then metal phytoextraction can be increased more than the plant can take up normally. Use of plant growth regulators (PGR) such as auxins and cytokinins have shown to enhance phytoremediation abilities of non-hyperaccumulating plants by increasing their growth and biomass. Enhanced bioavailability of iron by applying plant hormone indol-acetic acid (IAA) via a mechanism different from that involving siderophores has been reported. IAA is also produced by many plant growths promoting rhizobacteria (PGPR) such as *Pseudomonas* and *Acinetobacter* strains which result in enhanced uptake of iron, zinc, magnesium, calcium, potassium and phosphorus by crop plants. Further research needs to be carried out to find suitable combination of plant, PGPR, and soil type in order to investigate their potential(s) in increasing metal uptake by hyperaccumulator plants and improving the process of phytoextraction.

SUMMARY

Emissions of heavy metals to the environment occur via a wide range of processes and pathways i.e. to the soil (and hence into groundwater and crops), to surface waters (via runoff and releases from storage and transport) and to the air (during combustion, extraction and processing). Atmospheric emissions tend to be of greatest concern in terms of human health, both because of the quantities involved and the widespread dispersion and potential for exposure that often ensues. Recently, there has been increasing concern towards the health hazards associated with heavy metals entering the food chain via soils.

Various physico-chemical and biological remedial technologies have been developed. Physical and chemical methods of remediation of contaminated soils are mainly applicable to relatively small areas and are unsuitable for very large areas such as a typical mining site or industrially/agrochemically contaminated soils. Biochemical and plant-based phytoremediation technologies are applicable for removing metals from areas of low concentrations with shallow soils and water, although longer treatment times may be required.

However, it is a natural, clean, economic and environment friendly alternative to physical and chemical methods of clean up.

Further research is needed to investigate various chemical aspects of metal accumulation, e.g. diffusion of metals, release of specific chelators in the rhizosphere by accumulating and non-accumulating plants, the dynamics and persistence or decomposition of chelates and metal–chelate complexes in the rhizospheres/soil. This knowledge may enable us to understand other contributing soil and environmental processes. In addition, we need to understand the mechanisms involved in mobilization and transfer of metals in order to develop future strategies and optimize the phytoextraction process.

9

Arsenic Threat in Soil-Plant Environment

Arsenic (As) is a well-known toxic chemical that the Environmental Protection Agency (EPA) and the World Health Organization (WHO) list as a known carcinogen. Arsenic is found in a wide variety of chemical forms throughout the environment and can be readily transformed by microbes, changes in geochemical conditions, and other environmental processes.

THE ENVIRONMENTAL CHEMISTRY OF ARSENIC

Arsenic is commonly found throughout the environment in a wide array of chemical species that vary in toxicity and mobility. Many of these chemical species can be transformed due to biological activity or other changes in the environment, such as a change in oxidation-reduction potential and pH. This prospect for natural environmental change creates the possibility that a wide variety of arsenic species are constantly transforming at any time.

Arsenic in the environment is often associated with elements like sulphur (S), gold (Au), silver (Ag), copper (Cu) and uranium (U). Arsenic concentrations in environmental media are presented in Table 9.1. Arsenic occurs as a major constituent in more than 200 minerals, including elemental As, arsenides, sulphides, oxides, arsenates and arsenites. A list of some of the most common As minerals is given in Table 9.2. Arsenopyrite (FeAsS), realgar (As_2S_2) and orpiment (As_2S_3) are the most important of these minerals. They may be considered as As ores because of the high amount of As in these compounds.

Arsenic may be released to soil, surface water, groundwater, and the atmosphere from sulphide ores of other metals including Cu, Pb, Ag and Au. Where arsenopyrite is present in sulphide ores associated with sediment-hosted Au deposits, it tends to be the earliest-formed mineral, derived from hydrothermal solutions and formed at temperatures typically of 100°C or more.

Table 9.1. Arsenic concentrations in environmental media.

Environmental media	Arsenic concentration range
Air, ng/m^3	1.5 - 53
Rain from unpolluted ocean air, ug/L (ppb)	0.019
Rain from terrestrial air, ug/l	0.46
Rivers, ug/l	0.20 - 264
Lakes, ug/l	0.38 - 1000
Groundwater (well), ug/l	1.0 - 1000
Seawater, ug/l	0.15 - 6.0
Soil, mg/kg	0.1 - 1000
Stream/river sediment, mg/kg (ppm)	5.0 - 4000
Lake sediment, mg/kg	2.0 - 300
Igneous rock, mg/kg	0.3 - 113
Metamorphic rock, mg/kg	0.0 - 143
Sedimentary rock, mg/kg	0.1 - 490

Source: USEPA, 2000

The main species of arsenic found in the environment are the arsenic (III) and arsenic (V) oxyacids. In many environments, the arsenic (V) is often deprotonated as an arsenic (V) or arsenate anion; in contrast the arsenic (III) oxyacid remains in its neutral form as arsenite. Arsenate, arsenate anions, along with neutral arsenite constitutes the main targets for field analytical assays. In contaminated soils, inorganic arsenate is the predominant species. In general, the arsenate and other arsenic (V) species are immobilized on geologically available surfaces, usually as iron oxides. Although, arsenic (V) compounds are considered a low risk, bacterial and other environmental activities can readily convert them back into more mobile and more toxic forms of arsenic.

Table 9.2. Major arsenic minerals occurring in nature.

Mineral	Composition	Occurrence
Native arsenic	As	Hydrothermal veins
Niciolite	NiAs	Vein deposits and norites
Realgar	AsS	Vein deposits, often associated with orpiment, clays and limestones, also deposits from hot springs.
Orpiment	As_2S_3	Hydrothermal veins, hot springs, volcanic sublimation products.
Cobaltite	CoAsS	High-temperature deposits, metamorphic rocks.
Arsenopyrite	FeAsS	The most abundant As mineral, dominantly in mineral veins.
Tennantite	$(Cu, Fe)_{12}As_4S_{13}$	Hydrothermal veins.
Enargite	Cu_3AsS_4	Hydrothermal veins.
Arsenolite	As_2O_3	Secondary mineral formed by oxidation of arsenopyrite, native arsenic and other As minerals.
Claudetite	As_2O_3	Secondary mineral formed by oxidation of realgar, arsenopyrite and other As minerals.
Scorodite	$FeAsO_4.2H_2O$	Secondary mineral
Annabergite	$(Ni,Co)_3(AsO_4)_2.8H_2O$	Secondary mineral
Hoernesite	$Mg_3(AsO_4)_2.8H_2O$	Secondary mineral, smelter wastes.
Haematilite	$(Mn,Mg)_4Al(AsO_4)(OH)_8$	Secondary mineral
Conichalcite	$CaCu(AsO_4)(OH)$	Secondary mineral
Pharmacosiderite	$Fe_3(AsO_4)_2(OH)_3.5H_2O$	Oxidation product of arsenopyrite and other As minerals.

Groundwater and soil also contain organoarsenic species-monomethylarsenic acid, dimethylarsenic acid, trimethylarsine oxide, and trimethyl arsine. In general organoarsenic compounds are less toxic than their corresponding oxyacids. Although, usually found in lower concentrations, under the right conditions, they can be found in very high concentrations. In freshwater lakes, methylated arsenic can make up to 60% of the total arsenic. There are also arsenic sulphur species that constitute a sizable portion of arsenic geology and reducing environments in sediment and in solution. Although, all of these species are not as common or currently believed to be as toxic as arsenic oxyacids they constitute a sizable fraction of the naturally occurring arsenic.

Another important key to understanding the environmental risk from arsenic is bioavailability, defined as the measure of the amount of arsenic that can be absorbed by a living organism. Although, bioavailability is likely to play a strong role in future environmental regulatory decisions, it has not received widespread regulatory and public acceptance.

Arsenic in groundwater and its fate and transport in the environment have become matters of great concern in Bangladesh, India and several other countries. The increased soil content of arsenic is a result of the technogenic contamination due to widespread use of arsenic rich groundwater for irrigation. Soil contamination with arsenic may have toxic effect on vegetation and the animals feeding on the same. Arsenic may not be readily translocated from root to shoot, and different crop plants exhibit different tendencies to accumulate and tolerate arsenic. Besides, phytotoxicity due to increased arsenic in soil/water and its long-term impact on agricultural yield is another major concern.

ORIGIN OF ARSENIC IN SOIL

Arsenic (As) can be present in soils as a metalloid and as chemical compounds of both inorganic and organic forms. Arsenic can enter terrestrial and aquatic environments through both natural geologic processes (geogenic) and human (anthropogenic) activities. The natural pool of As in surface soils arose from the net of geological, hydrological and soil-forming biogeochemical processes. Under typical soil-forming conditions, the nature of soil As is controlled by the lithology of the parent rock materials, volcanic activity, weathering history, transport, sorption, biological activity and precipitation. Since As accumulates due to weathering and translocation in colloid fractions, its concentration is usually higher in soils than in parent rocks. The national academy of sciences, medical and biological effects of environmental pollutants, established the average of As content of the Earth's crust as 2.5 mg/kg. Arsenic is also commonly associated with sulphides, e.g. in sulphidic ore deposits. Other natural sources of As include volcanic activities, wind-born soil particles, sea salt sprays and microbial volatilisation of As. Arsenic is also a typical component of the thermal waters, its concentration can reach up to 47 mg/l in carbonaceous chloride mineral springs.

The primary anthropogenic contributions of As to soils are from the combustion of municipal solid waste, application of arsenical pesticides (herbicides, fungicides and insecticides), land application of solid waste/sewage sludge, river and irrigation waters, mining and smelting of As-containing ores, combustion of fossil fuels (especially coal), land filling of industrial wastes, release or disposal of chemical warfare agents, manufacturing of metals and alloys, petroleum refining, and pharmaceutical manufacturing. Another potential contributing source of As in soil is the current use of chromated copper arsenate (CCA) as wood preservative. Most agricultural uses of As have been banned, however, sodium salts of methylarsonic acid are still used in cotton fields as herbicides. Organic As is also a constituent of feed additives for poultry and swine for the control of coccidian intestinal parasites and to improve feed efficiency, and appears to concentrate in the resultant animal wastes. There is a significant use of As in the production of lead-acid batteries, while small amounts of very pure As metal are used to produce gallium arsenide, which is a semiconductor used in computers and other electronic applications.

TOXICITY OF ARSENIC IN SOILS

Important factors affecting As chemistry, and its mobility in soils, are soil solution chemistry (pH and redox conditions), solid composition, As-bearing phases, adsorption and desorption, and biological transformations, volatilisation, and cycling of As in soil. Metal sulphide concentrations, temperature, salinity and distribution and composition of biota appear to be significant factors that determine the fate and transport of As.

The solubility, mobility, bioavailability and hence toxicity of arsenic (As) in soil-crop system depends on its chemical form, primarily the oxidation state. The inorganic forms of arsenic in soils and sediments include arsenites and arsenates. The organic forms, namely dimethyl arsinic acid or cacodylic acid, which on reduction (e.g., in anoxic soil conditions) forms di- and trimethyl arsines, are also present. Under oxidizing aerobic conditions (at high E_h), arsenic acid species ($H_3AsO_4^o$, $H_2AsO_4^-$, $HAsO_4^{2-}$, AsO_4^{3-}) are stable in soil environment, whereas under mildly reducing conditions (such as one encounters in flooded soils), arsenous acid species ($H_3AsO_3^0$, $H_2AsO_3^-$, $HAsO_3^{2-}$) are the stable forms. Furthermore, it has been recognized that As (III) is more prevalent in soils of neutral pH range (and in most groundwater), as in the soils of the affected belt of West Bengal, India and Bangladesh, than otherwise thought, and hence is of concern. This is primarily because As (III) exists as a neutral, uncharged molecule, $H_3AsO_3^0$ ($pK_a = 9.2$), at the pH of neutral soils and most natural groundwater. The latter is thus more mobile in being less strongly retained by most mineral surfaces. Arsenic species in soil environment are subject to both chemically and microbially mediated oxidation or reduction and methylation reactions (Fig. 9.1).

In course of several attempts to fractionate soil arsenic in different fractions, the 'freely exchangeable' arsenic (anion exchange resin–extractable As), rather than total arsenic in soil, has been found to provide a satisfactory index of potential bioavailability and mobility, and hence toxicity of As in soil.

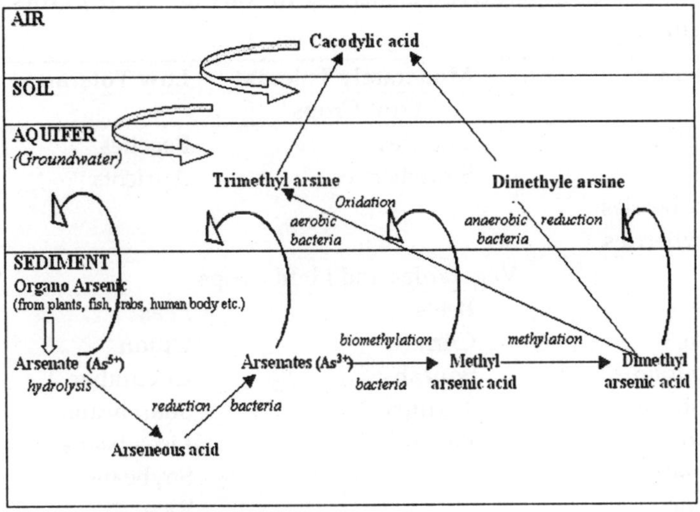

Figure 9.1. Microbially mediated oxidation-reduction and methylation reactions in soil environment

The biological availability and phytotoxicity of arsenic in soil increases on reduction to As (III) which is facilitated on flooding the soil, leading to fall in E_h and an increase of pH. Under such conditions, arsenic lost from soil solution could be accounted for primarily as surface-bound with some losses of gaseous arsines.

BIOACCUMULATION OF ARSENIC IN FOOD CROPS

There are reports on arsenic in vegetables and the dietary intake of As in India, Bangladesh and many other countries. As per findings of arsenic polluted areas in West Bengal, high concentrations of As were detected in vegetables and cereals. The reason for the high arsenic concentration in vegetables and other foodstuffs is the use of arsenic contaminated water for irrigation. Due to the constant use of As-contaminated irrigation water, soils were also contaminated by arsenic. Absorption of arsenic is influenced by the concentration of As in soils and the nature of plant species. A high concentration of arsenic is reported in rice which is a staple food in Asian countries. Sensitivity of various crop plants to arsenic is given in Table 9.3.

Table 9.3. Comparative sensitivity of various crop plants to arsenic.

Tolerant	Moderately Tolerant	Low Tolerance
	Fruit Crops	
Apples	Cherries	Peaches
Grapes	Strawberries	Apricots
Raspberries		
Dewberries		
	Vegetables and Field Crops	
Rye	Beets	Peas
Mint	Corn	Onion
Asparagus	Squash	Cucumber
Cabbage	Turnips	Snap beans
Carrots	Radish	Lima beans
Tomato		Soybeans
Potato		Rice
Wheat		Spinach
Oats		
Cotton		
Peanuts		
Tobacco		
	Forage Crops	
Sudan grass	Crested wheat grass	Alfalfa
Bluegrass	Timothy	Bromegrass
Ryegrass		Clover
Meadow fescute		Vetch

Arsenic is accumulated mainly in the root system, to a lesser extent in the overgrown organs, inhibits the growth and fresh and dry biomass accumulation and causes physiological disorders, as well as reduction of the crop productivity. To some extent, its effect is due to the suppression of the high-affinity phosphate/arsenic uptake system. Arsenic acts as a phosphate analogue and is transported across the plasma membrane via phosphate transport systems. Once inside the cytoplasm, it competes with phosphate. For example, it replaces phosphate in ATP to form unstable ADP-As, and leads to the disruption of energy flows in cells. Arsenic also caused a reduction of the photosynthesis rate and damage the chloroplast membrane and disorganized the membrane structure.

Availability of arsenic for its uptake by plants is affected by several factors, such as the source, chemical speciation and soil parameters (pH, EC, organic matter, and colloid contents, soil texture and drainage conditions).

ARSENIC ACCUMULATING PLANTS/MICROBIAL SPECIES

A scan of literature reveals a number of plant/microbial species, known for arsenic accumulation/bioindicator, which can effectively remove arsenic (and other heavy metals) from the aquatic system, for instance, to the tune of 170 and 340 mg As/g dry weight of water hyacinth in its stem and leaves, respectively, when grown in a pond containing 10 mg As/dm^3. However, such accumulated arsenic in water hyacinth *(Eichhornia crassipes)* is also liable to leaching out in the water body, particularly so on decomposition of such aquatic weed. Consequently, appropriate precaution has to be exercised while interpreting the arsenic status of aquatic environment by water hyacinth accumulation. Hydrilla (*Hydrilla verticillata* Casp.) is another such aquatic plant. Pointed gourd, a vegetable creeper plant (*Trichosanthes dioica*), has also been found to accumulate arsenic when cultivated in the contaminated soils of West Bengal. Table 9.4 provides an overview on reported shoot, root and substrate concentrations of As hyperaccumulators and As tolerant plants.

The reported hyperaccumulation of arsenic from the contaminated soils by the brake-fern (*Pteris vittata*), and its subsequent translocation into the above-ground biomass suggests that the plant-accumulated arsenic is present almost entirely in the toxic inorganic forms, with the proportion of highly toxic As (III) being, in fact, much greater in the plant body than that of the less toxic As (V) form, as compared to the distribution of these two forms in the contaminated soil in which the fern grows. Conversion of the plant-accumulated inorganic forms of the toxin to non (or less)–toxic organometallic forms by plant metabolism would certainly aid the detoxification process. Such detoxification within the plant body assumes importance, particularly in view of the report that arsenic in plant residues may be mobilized by a unicellular alga, namely *Polyphysa peniculas*, resident in normal (moist) agricultural soils.

Table 9.4. As concentrations in hyperaccumulator and As tolerant plants.

Plant species	As in plants (mg/kg) Frond/shoot	Root	As in soil (mg/kg)
Hyperaccumulators			
Pteris vittata	755 - 22630	303	6 - 1500
Pteris calomelanos	8000	88	135
Tolerant plants (non-accumulators)			
Agrostis capillaries	3470	--	26500
Agrostis catellana	170	1000	1700
Agrostis delicatula	300	1800	1700
Cynodon dactylon	1600	10850	9530
Paspalum tuberosum	1130	--	7670
Spergularia grandis	1175	--	7670

Source: Fitz and Wenzel, 2002.

Several other crop plant species (rice, elephant-foot yam, green gram, etc.) are also reported to accumulate As in substantial quantities. The toxicity of arsenic species in plant body is reported to follow the order AsH_3>As(III)>As(V)>MMA (monomethylarsonic acid)>DMMA (dimethylarsinic acid).

Arsenate tolerance by the grassy weeds, namely *Agrostis castellana* and *A. delicatula*, has been discussed in terms of the comparison of the corresponding reduction of maximum root growth (MRG) with that in the sensitive populations upon exposure to arsenic. These findings tend to suggest that most, if not all, cropped plant species irrigated with As contaminated water, or such as those cultivated on soils, or in water bodies containing high levels of the toxic metallocid, tend to accumulate arsenic.

A number of microbial species (e.g., the bacterial species, namely *Proteus* sp., *Escherichia coli, Flavobacterium* sp., *Corynebacterium* sp., and *Pseudomonas* sp.; the fungus, namely *Candida humicola*; the freshwater algae, namely *Chlorella ovalis, Phaepdactuylum tricornutum, Oscillatoria rubescens*) have been reported to possess varying degrees of arsenic accumulating abilities. However, the extent of transformation of species and detoxification of such accumulated arsenic is varied, and also to an extent, uncertain.

SUMMARY

Arsenic still ranked first in the list of 20 hazardous substances by the agency for toxic substances and diseases Registry. Arsenic, from both anthropogenic and geologic sources, has been considered one of the most toxic elements, affecting millions of people around the world. The source, uptake and release of arsenic in ground water is now established, the chemistry of arsenic toxicity is now well develop, and arsenic cycle is now a reality. Arsenic which is now harmful to both soil and plants and affects human in two ways, first through drinking water and letter through food. These joint actions accelerate the development of toxicity symptoms, which are of various forms and easily visible on skin and on some other parts. A prolongs neglect may leads to development of several forms of cancer. The avoidance of use of arsenic rich irrigation water, agrochemicals containing As; reliable and simple methods for soil/water arsenic testing in field or laboratory, that would allow decision making for remediation; and an adequate disposal plan in the case of phytoremediation are some of the issues that need urgent attention.

10

Acid Mine Drainage and Soil Environment

Acid Mine Drainage (AMD) is the term used to describe 'leachate' that has become acid due to oxidation of sulphur containing minerals exposed to air and water. The AMD problem occurs as a result of coal mining activities, worldwide and is presently a very severe environmental problem due to increased coal mining activity.

Acid drainage is produced wherever a mine of any type in permeable formation intersects with the water table, aquifer, and perched water body or surface water finds its way into a mine. In those terrains, where sulphides (particularly pyrites) are present in the ore or country rock AMD may occur. Oxidation of pyrite primarily by the chemoautotroph *Thiobacillus ferroxidans*, produce sulphuric acid during mining releasing large quantities of acid resulting in the acidification of waterways. Acid drainage may be recognized by orange or yellow ferric hydroxide precipitate in streambeds and / or a sulphurous smell of water. Acidic nature of soil may be a possible indication to potential acid drainage problem.

Once established, acid drainage is often difficult and costly to treat. Management by prevention requires characterization of overburden or waste material and knowledge of the hydrology of the site, so that the likely occurrence of acid drainage can be predicted and potentially acid producing material selectively handled and isolated. Where the potential for acid drainage exists, provision in the planning stage to prevent acid drainage is essential.

CHARACTERISTIC OF ACID MINE DRAINAGE

1. Acid mine drainage can produce pH values < 2.5 (strongly acidic).
2. Strongly acidic pH of the AMD dissolve soil minerals and produce toxic concentration of Fe, Al, Mn and trace or heavy metals such as Cu, Cd, Ni, Cr, As, Zn, Hg, Pb, etc. and result in high salt levels.
3. The water is characterized with low conductivities and reflects the non–availability of calcareous material and thus unable to afford the natural buffering capacity in the geologic regimes.
4. The strongly acidic environment due to AMD is hostile to the growth of microorganisms, other soil fauna and flora and higher plants.

5. The flushes of acidity from mine spoil adversely affect crops growing on adjacent better lands and have been implicated in the death of aquatic plants, shell fish, crustacean and fish in aquatic ecosystem.

Table 10.1. USEPA Classification of Acid Mine drainage.

Parameter	Class 1 Acid Drainage	Class 2 Partially Oxidised and or Neutralised	Class 3 Oxidised and neutralized and/or alkaline	Class 4 Neutralised and not oxidized
pH	2.0- 4.5	3.5-6.6	6.5-8.5	6.5-8.5
Acidity (mg/l)	1000-15000	0-1000	0	0
Ferrous iron (mg/l)	500-1000	0- 500	0	50-1000
Ferric iron (mg/l)	0	0-1000	0	0
Aluminium (mg/l)	0-2000	0-20	0	0
Sulphate (mg/l)	1000-20000	500-10000	500-10000	500-10000

The United States Environment Protection Agency (USEPA) has formulated the AMD classification as given in Table 10.1.

THE MECHANISM OF FORMATION OF AMD

Sulphur in coal (and other metalliferrous ore bodies) gives rise to the origin of AMD. Distribution of sulphur forms is not the same in all the coals. In fact, in most of the coals, the pyritic sulphur content is predominant. The pyritic sulphur content of coal varies from coal seam to coal seam and varies from one geographic region to another. Coals are arbitrarily classified as low, medium and high sulphur (1% or less, 1-3% and >3% respectively) content coal. There are extreme variations in the total pyritic and organic sulphur contents. Sulphur as such does not present at significant concentration but occurs in inorganic (sulphate and sulphide forms) and organic combinations. However, organic sulphur is structurally bound and inert and hence does not produce acidity in mine drainage to a considerable extent.

The potential of a coal seam to produce acidic drainage is governed by many factors. Significant among them are –

(a) Relation to land surface

(b) Water table

(c) Climatic pattern

(d) Hydrological pattern

(e) Geology of the area

It is generally assumed that the amount of acid produced by chemical weathering of an exposed coal seam is almost totally due to oxidation of pyrite, and is therefore not a direct function of the total sulphur contained in the seam. But, as the total sulphur increases, the poly-sulphide content also increases and hence the potentiality. Since organic sulphur is structurally bound in coal, it cannot be washed away from a seam during chemical weathering. Quantities of sulphate sulphur and free sulphate present are too insignificant to produce much acidity in streams draining mine areas. This leaves only pyritic sulphur as the primary source of AMD formations.

Pyritic sulphur can be found in many structural parts of the coal-mine and adjacent rock beds. The binders or partings are shale layers that overlay or underlay the horizontal coal seam and often contain pyrite, clay, calcite and other minerals. These balls may be distributed randomly throughout the bed or they may be concentrated in layers. Lenses (circular or elliptical masses of pyrite usually about 30 cm in diameter and several centimeters thick), veins or plates are thin streaks of pyrite in the form of finely dispersed particles ranging between a fraction of a micron and 50 micron in diameter. The most common pyritic materials usually associated with the acid formation are large sulphur balls present in coal strata.

PREVENTION TECHNOLOGY OF AMD

The most effective and economical method of controlling acid drainage is to prevent its formation. Thus, treatment procedure varies according to site conditions. Oxygen and water are necessary to initiate acid formation and therefore prevention methods aim to exclude either reactant from the pyritic material. This involves controlled placement of acid forming materials and appropriate water management strategies.

Prevention is dependent on identifying the pyritic material before mining in order to adopt mining procedure that is able to selectively handle acid forming material. If calcareous strata or other alkaline material, which can neutralize any acidity generated, are available mining methods should enable blending of material within the mining area. Some of the methods for preventing AMD are given below:

a) Control the hydrology of the site to prevent water from contacting pyritic material by diverting surface water away from pyritic material and preventing and subsequent infiltration.

b) Isolate the pyritic material from water by 'capping' with clay or impermeable materials. The cap can then be covered with soil and vegetation established. This technique reduces infiltration and leaching. In practice, it is very difficult to completely isolate pyritic spoils from water or air contact. Unlike tailing storage areas, waste dumps are unlikely to have an impermeable or semi-permeable base or sides. The task of reshaping and encapsulation is consequently greater and more costly.

c) *Submerge the acid forming material.* This can be an effective strategy where sufficient water is available. A water cover sufficient to maintain the partial pressure of oxygen below 1% is necessary to inhibit pyrite oxidation.

d) *Sealing of mine portals.* This can reduce the AMD problem by reducing water and air infiltration. The portals are typically sealed with a plug of expansive grout with steel reinforcement. A 'wet seal' is provided on the sealed wall, which includes pipe drains through the grout to collect acidic water from the mine. Water should be treated in laboratory for analysis from time to time. Sudden and dangerous bow outs due to failure of wall, erected for sealing, may pose problem in such treatments.

e) *Deep well injection or grout injection.* This treatment involves the injection of a grout (typically a mixture involving fly ash) into a mine to control AMD. The grout must be designed for proper flowability, chemical stability and compressive strength properties and is typically through holes drilled on 50 X 200ft centres depending on the mine condition.

f) *Mine Capping.* Capping can prevent or reduce rainfall from reaching acid forming units in an abandoned mining area or in backfilled mine. Capping is generally used in surface mines. The cap is nothing but spreading of a layer of preserved topsoil mixed with fly ash (alkaline) and over this layer seeding can be done. For capping to be effective, horizontal components of groundwater must be negligible.

g) Dilute acid-mine drainage to an acceptable effluent quality and employ standard wastewater treatment method for neutralization and removal of dissolved solids. The technique includes lime or limestone treatment accompanied by aeration or oxidation process to convert ferrous iron to ferric iron; neutralization with soda ash, caustic soda and anhydrous ammonia; reverse osmosis; ion exchange; electro-dialysis; evaporation; ozone oxidation; desulphating; sulphide iron removal; microbiological control and permanent iron removal.

h) *Incorporation of lime or other neutralizing materials.* Treatment of acidity for surface areas / mined out areas or waste dumps in mines require use of chemicals. The chemicals usually used for Acid Mine Drainage treatment include limestone, hydrated lime, soda ash, caustic soda etc. Injection of neutralizing fluids e.g. sodium carbonate, anhydrous ammonia or caustic soda into mine dumps to intercept flow paths of acid drainage is also an approach for treatment. Neutralizing capacity of the available material and the lime demand of the mined out areas should be tested to determine feasibility. Some wastes such as 'Calcium-silicate Slag' and 'Alkaline Flyash' is also useful as neutralizing agent. Care should be taken to prevent toxicity problems caused due to other trace elements in such neutralizing agent.

i) *Channeling runoff.* From the damaged area or dump site to selected recharge areas a ditch filled with alkaline material is made. This is called Anoxic Limestone Drain (or ALD). Thus, an ALD is an adequately sized buried channel containing limestone that is designed to limit diffusion of atmospheric oxygen with mine discharge.

j) *Inline aeration.* Collection of acid drainage down stream for chemical treatment and treating with aeration techniques.

k) *Evaporation.* In areas where evaporation consistently exceeds precipitation, disposal by evaporation may be a feasible solution for AMD control. Safe disposal of sludge with elevated levels of heavy metals and salt is then required.

l) *Biological treatment.* Since, AMD is caused due to bacterial activity controlling the bacteria (*Thiobacillus ferrooxidans)* in situations where it plays a significant role in accelerating acid formation is a one way of treating AMD. Research studies have proved that 'selective bacteria type' when implanted in affected soil can eat soil acidity in a natural way.

m) *Constructed wetlands.* Directing acid drainage to artificial wetlands where biological production of bicarbonate neutralizes the acidic drainage. Metals are removed through hydrolysis and biological formation of insoluble sulphides and carbonates. Constructed wetlands utilize soil and water-borne microbes associated with wetland plants to remove dissolved metals from mine drainage. Unlike chemical treatment, however, wetlands are passive systems requiring little or no continuing maintenance. This is a relatively new treatment technology with many specific mechanisms. Wetlands are generally more effective in removing iron than manganese which causes AMD. The greatest utility of wetlands appears to be in the treatment of small flows of a few gallons per minute.

n) *Mycorrhizal fungi* are an important and essential component in re-establishing plant communities on AMD affected sites. Vesicular arbiscular (VAM) and ectomycorrhizal are the most common. *Pisolithus tinctorus* is dominant mycorrhizal partner in the mine spoils to which it is admirably adapted.

SOIL FERTILITY IN MINE SPOILS

Nitrogen and phosphorus are most limiting growth factors in mine spoils. Phosphorus is a major limited nutrient during the colonization and early succession process on surface-mined lands. Lack of mineralized organic N and lower mineralization rates affect the availability of N to plants in mine spoils. Available carbon source is a critical factor to stimulate spoil micro-flora in mine spoils.

Use of fertilizer P before seeding corrects P deficiency of mine spoils. The ferrous ion, released during oxidation of pyrite with atmospheric oxygen and water, is immediately converted into ferric phosphate as phosphate ions in solution have a great affinity for metallic ions and thus prevent formation of ferric sulphate and thereby initiation of further oxidation of pyrites by ferric ions:

$$3Fe + 2PO_4 \rightarrow Fe_3\,(PO_4)_2$$

A large pool of organic N and a high rate of ammonification are necessary to sustain vegetation and to prevent N immobilization. High levels of organic matter in mine spoils improve aggregation and infiltration capacities and increase the availability of nutrients. Nitrogen-fixing legumes or annual fertilization with N is essential to build up a pool of mineralized organic N in mine spoils. Sludges, paper mill waste, sawdust, fly-ash, mulches and other amendments are used as an alternative to fertilizers.

Mine spoils can be stabilized through the establishment of beneficial plant cover. Seeded species, initial seeding success, cultural practice and weather influence plant succession of mine spoils. Seeding or planting of plant species also controls soil erosion and increased plant diversity. Many plant communities are highly productive for 2-5 years but productivity and vigour declines due to large build up of litter which ties up substantial amount of C and N later on. Well adapted species have been recommended to establish self-sustaining cover which required little or no-maintenance activities.

SUMMARY

The adverse environmental effects of AMD are many and potentially serious. Degradation of agricultural and non-agricultural land, decline in agricultural productivity, adverse effect on biodiversity, pollution of land and water are some of the visible environmental effects of acid mine drainage (AMD). Objectionable features of AMD are low pH and high levels of sulphur and iron. Presence of AMD depletes essential plant nutrients in soil and oxygen level in water. This in turn increases acidity and toxicity as well. Acute acidity due to acid mine drainage, nutrient deficiency, sandy nature of the spoil increases in soil erosion and moisture stress characterize the mine spoil.

Adequate care needs to be taken to prevent the generation of AMD at source and minimize land subsidence after mining. This can be done by refilling of the excavated mines with mine spoils or sub-soils. The seepage of AMD water from the mined areas should be properly channelised to stop deterioration of soil and water quality in agricultural fields and low lying marginal lands.

The quality of land and water, flora and fauna of the mine affected area needs to be improved subsequent to the initial eco-restoration works. Immediate measures should be taken to check soil and water conservation, to improve land quality by using additives like fertilizers, manure, lime material, plantation of appropriate local plant species, etc. The treatment of AMD by a blue-green algae-microbial consortium, containing powdered manure, wood chip and soil are cost effective reclamation solutions. Practices must be adopted for long-term improvement for land quality.

11

Agrochemicals in the Soil Environment

Agrochemicals have made important contributions to the successes of the "green revolution." This has helped to increase the food supply for the rapidly increasing population of humans on Earth. However, the use of certain agrochemicals has also been associated with some important environmental and ecological damages. Excessive use of fertilizers, for example, can lead to the contamination of groundwater with nitrate, rendering it unfit for consumption by humans or livestock. In addition, the run-off of agricultural fertilizer into streams, lakes, and other surface waters can cause an increased productivity of those aquatic ecosystems, a problem known as eutrophication. Environmental concern associated with fertilizer use already discussed in the earlier chapter. This chapter, however, deals with pesticides.

The pesticides become a major agrochemical inputs in intensified agriculture and can have considerable impact on soil environment. There is a widespread, even global contamination of the environment with some types of persistent pesticides, especially with organochlorines such as DDT, dieldrin, and aldrin. Most modern usage of pesticides involves chemicals that are less persistent than DDT and related chlorinated hydrocarbons. However, severe damages are still caused by the use of some newer pesticides.

TRANSFORMATION OF PESTICIDES IN THE SOIL

Soil is the ultimate sink for bulk of the pesticides applied either to the soil or to the foliage. According to several estimates, as much as 50 per cent of the pesticides applied to the foliage eventually reach the soil. A pesticide reaching the soil is acted upon by several forces i.e. physical, chemical and biological-independently or in combination. The most extensive degradation of even complex organic molecules often to inorganic end products, is achieved through biological means. But in non-biological reactions the degradation is less extensive and far from complete.

Retention refers to the ability of the soil to hold a pesticide in place and not allow it to be transported. Adsorption is the primary process of how the soil retains a pesticide and is defined as the accumulation of a pesticide on the soil particle surfaces. Pesticide adsorption to soil depends on both the chemical properties of the pesticide (i.e., water solubility, polarity) and properties of the soil (i.e., organic matter and clay contents, pH, surface charge characteristics, permeability). For most pesticides, organic matter is the most important soil property controlling the degree of adsorption. For most pesticides, the degree of adsorption is described by an adsorption distribution coefficient (Kd), which is mathematically defined as the amount of pesticide in soil solution divided by the amount adsorbed to the soil.

The transformation of pesticides added to soil depends upon adsorption or retention on to clay and organic matter, leaching, volatilization, uptake by soil organisms and plants, movement with runoff water, microbial and chemical degradation and photolysis.

PERSISTENCE OF PESTICIDES IN SOILS

Persistence is relative terms, time taken for 75 to 100% loss of pesticide is normal arable soil under recommended practices of application is taken as persistence value. The highly immobile compounds of pesticides are phorate, parathion, ethion, zineb, benomyl, paraquat, trifluralin, heptachlor, endrin, aldrin, chlordane, toxaphene, DDT, etc. Some of these are already banned for use in agriculture. The commonly used herbicides such as attrazine, alachlor, propachlor, simazine, propanil, diuren, etc. have moderate to high immobility indicating greater persistence in soil.

When the pesticides are used repeatedly in each crop season and those applied to soils which are poor in organic matter and microbial biodiversity, the chemicals are bound to accumulate for longer periods of time causing much environmental pollution and yield depressions. Organochlorines are very persistent, typical values are - chlordane 5 years, DDT 4 years, BHC 3 years, heptachlor, aldrin and dieldrin 1-3 years. Persistence of organophosphorous and carbamate insecticides is low - diaznon 3 months, phorate 2 weeks and malathion and parathion 2 weeks. Persistence of herbicides varies e.g. propazine and picloram 18 months, simazine 12 months, atrazine 10 months, trifluralin 6 months, 2,4,5-T 5 months and 2,4-D 1 month. Most fungicides remain only for a few months in soils and most of the fumigants disappear in 3 to 4 weeks. Fungicides Dodin persist in soil for more than 45 days but it have marginal effect on soil population and no effect on soil dehydrogenase activity.

Factors Influencing Pesticide Persistence

Factors influencing the rate of degradation of a pesticide in the soil environment are many and interrelated. The important factors affecting pesticide persistence are discussed here.

Soil moisture content. A pesticide is degraded faster in moist soil than in dry soil. However, the degradation rate may rise or fall after further increase in the water content to saturation, as in flooded soils. The degradation of certain chlorinated insecticides such as BHC and endrin proceeded more rapidly under flooded conditions than under non-flooded conditions. Alternate flooding and drying helps in more extensive degradation of pesticides because not all organic molecules undergo decomposition in flooded soils.

Organic matter. It increases the microbial activity in a soil and the availability of organic molecules to degradation process through sorption. Addition of rice straw or other organic matter to a flooded soil accelerated the degradation of isomers of BHC, endrin and carbofuran.

Redox Potential. It has marked effects on the persistence of pesticides degraded by oxidative or reductive processes. For example, degradation of b - and g -isomers of BHC occur only in soils capable of attaining redox potential of -40 to -100 mV within a few days after flooding. The degradation of b –BHC can be only at a potential lower than that required for the degradation of g –BHC. This explains the long persistence of b –BHC in the environment. The drop in redox potential of soils upon flooding can be further hastened in the presence of organic matters besides microbial activity.

Sorption and desorption. This regulates the amounts of pesticides available for losses through degradation (chemical and microbiological) and transport (volatilization, leaching, surface run-off). Organic matter is the most important single factor affecting the sorption process in soil. Inorganic soil constituents have also been implicated in the sorption of the pesticides, especially in soils with organic matter (> 2 %) and in soils oxidized with H_2O_2.

Mineral constituents. Pesticides sorbed by inorganic soil constituents can undergo surface catalytic decomposition. Inorganic soil constituents notably clay minerals and free iron oxides participate in the sorption pesticides by aerobic soils. Besides catalytic decomposition of pesticides after sorption by minerals, interaction between the pesticides and the metabolic end products of minerals also takes place.

EFFECT OF PESTICIDES ON SOIL MICROBES

Most of the fungicides, nematicides and insecticides applied to soil suppress temporarily the microbial activity, and regain their activity soon. Addition of organic matter helps to reduce toxicity due to adsorption as well as greater breakdown due to enhanced microbial population. Pesticide concentrations in excess of recommended doses usually reduce the soil respiration.

Pesticides have adverse impact on soil fauna affecting the populations of arthropoda (insects, mites, spiders, millipedes and centipeds), mollusca (snails, slugs), annelids (earthworms) and protozoans (paramecium, amoeba), etc. which are very important in maintaining soil structure and nutrition. Earthworms concentrate DDT manifold and show higher respiratory metabolism. Populations of predacious and parasite insects in soil are reduced by repeated application of pesticides, especially organochlorines. Microorganisms are killed not only from soil applied pesticides but also from drift from arial sprays or washed off from foliage.

Of all biochemical processes, nitrification is most sensitive to pesticides. It was found that dieldrin and heptachlor suppressed nitrification up to 8 weeks and after 16 weeks toxic effects disappeared completely. There is nitrate accumulation showing specific inhibitory effect on *Nitrobacter*. The inhibition of *Nitrosomonas* led to accumulation of ammonium.

PESTICIDES DEGRADATION

Microorganisms are more important in degradation of pesticides than physico-chemical mechanisms. Basically 5 processes are involved in the microbial transformations.

a) *Biodegradation* in which pesticide act as a substrate for growth.

b) *Co-metabolism* which is most prevalent and in which pesticide is transformed but does not serve as a substrate for growth.

c) *Polymerization* in which pesticide molecules are linked together with other pesticides or naturally occurring compounds. Microbial polymerization plays a significant role in incorporation of xenobiotics into organic matter.

d) *Accumulation* in which pesticide is incorporated into the microorganisms. Dead cells absorb more insecticides than living cells. Thus it also enters into food chains of fishes, etc. in aquatic environments. Dead cells reach sediments where the pesticide molecules can be re-released posing a delayed pollution problem.

e) *Secondary effects of microbial activity* in which microbial activities contribute to alteration of environmental parameters such as pH, Eh, etc. that are conducive to secondary or non-enzymatic transformations of pesticide molecules.

Many pesticides are degraded by reductive reactions that proceed non-enzymatically under anaerobic conditions, e.g. DDT, methoxychlor and heptachlor breakdown.

The idea of using microorganisms or their enzymes to degrade pesticides has been well documented. For example, degradation of DDT by inoculation of *Aerobactor aerogenes, Pseudomonas stutzeri* and *P. aeruginosa* for parathion, etc. Alternatively pesticides breakdown can be accelerated by enriching the microbes already present, e.g. 2,4-D mineralization can be stimulated by addition of sachhrides and other organic substances. However, there are limitations to microbial application due to soil conditions, substrate availability and preference. Use of enzyme are more advantageous as they can withstand at higher concentration of salts, solvent and temperature than microorganisms.

Genetic engineering is another possibility. During the last 20 years numerous genes in bacteria coding for enzymes involved in pesticide degradation have been shown to be plasmid encoded e.g. 2,4-D, genes coding for dehalogenases, parathion, DDT, kelthane, 3-chlorobenzoate, etc. using recombinant DNA techniques. It is possible to produce soil microbial populations that carry plasmids conferring ability to degrade recalcitrant pesticide molecules such as 2,4,5-T (*P. cepacia*) and its potent contaminant 2,3,7,8-tetrachloro-di-benzo-para-dioxin. The potential application of genetically engineered strains is still somewhat speculative.

SUMMARY

The capacity of the soil to filter, buffer, degrade, immobilize, and detoxify pesticides is a function or quality of the soil. The presence and bio-availability of pesticides in soil can adversely impact human and animal health, and beneficial plants and soil organisms. Pesticides can move off-site contaminating surface and groundwater and possibly causing adverse impacts on aquatic ecosystems also.

Researchers are constantly searching for non-chemical ways of dealing with many of these agricultural needs. Much attention is being paid to developing biological methods (using microorganisms) of enhancing soil fertility and dealing with pests. Unfortunately, economically effective alternatives to most uses of agrochemicals have not yet been discovered. Consequently, modern agricultural industries will continue to rely heavily on the use of agrochemicals to manage their problems of fertility, soil quality, and pests.

12

Gaseous Emissions from Soils

Gaseous emissions of oxides of nitrogen, CO_2, CH_4, H_2S, cholorofluorocarbon, etc. which cause global warming and thus has potentially dangerous consequences for agriculture and environment. On a global basis, atmospheric methane arises from bacterial decomposition of organic matter in lakes, ponds, marshes, rice fields and swamps. Annual emissions of methane are of the order of 500 Tg. Methane has 20 times greater radiative forcing than CO_2. There is increasing concern about emissions of methane from rice fields. The concentration of CO_2 has increased from 280 ppm in pre-industrial age to approximately 350 ppm at present and likely to increase to 460 ppm by 2030 and 555 ppm by 2100 AD. Oxides of nitrogen is 300 times more potent than CO_2, so even small emissions of this non-CO_2 gas can make a considerable contribution to global warming.

EMISSION OF OXIDES OF NITROGEN

Soils act as both sources and sinks of N_2O and NO. However, on the global scale the N_2O and NO emissions dominate the sink activity. The production and consumption of N_2O and NO in soils involves both biotic and abiotic processes. Numerous groups of microorganisms contribute to the production and consumption of NO and N_2O, but biological nitrification and denitrification are the dominant processes involved.

Biological denitrification is the reduction of nitrate (NO_3^-) or nitrite (NO_2^-) to gaseous N oxides and molecular N_2 by essentially aerobic bacteria. Nitrification is the biological oxidation of ammonium (NH_4^+) to NO_2^- or NO_3^- under aerobic conditions. Under oxygen limited conditions nitrifiers can use

NO_2^- as a terminal electron acceptor to avoid accumulation of the toxic NO_2^-, whereby N_2O and NO are produced.

The bacterial processes of denitrification and nitrification are the dominant sources of N_2O and NO in most systems. Only denitrification is recognized as a significant biological consumptive fate for N_2O and NO. The chemical decomposition of HNO_2 (or chemical denitrification), that is, the reduction of NO_2^- by chemical reductants under oxygen limited conditions and at low pH, can also produce N_2, N_2O and NO. Chemical denitrification generally occurs when NO_2^- accumulates under oxygen limited conditions, which may occur when nitrification rates are high, e.g. after application of NH_4^+-based mineral fertilizers or animal manure. This process may account for 15-20 percent of NO formation.

Plant uptake of NO and NO_2 is a further important biological process determining the exchange of NO_x at the Earth's surface. Some other abiotic processes may also produce and consume trace amounts of NO.

SOIL ENVIRONMENT FACTORS

Gas production and exchange with the atmosphere depend on N availability, denitrification and nitrification rate, soil moisture and temperature, soil pH, gas diffusion etc. In addition to soil and climate characteristics, agricultural management may also interfere with soil processes and thus influence N availability and the relative rates of N_2O and NO.

N availability. In agricultural soils, N inputs from various sources determine the amount of N flowing through the soil. N fertilizer, N inputs in agricultural soils, animal manure, crop residues, deposition and biological N fixation are strong stimulants of denitrification and associated N oxide fluxes. The magnitude of NH_3 volatilization also determines the availability of N for nitrification and denitrification. Studies have found denitrification to be lower when NH_3 losses are high in both upland soils and wetland rice systems.

Nitrification and denitrification rates. Whether denitrification or nitrification dominates depends on many different factors.

Nitrification is a relatively constant process across ecosystems, whereas denitrification rates are temporally and spatially variable. In most soils, availability of NH_4^+ and oxygen are the most important factors controlling nitrification. The major controls on biological denitrification include the availability of organic carbon, oxygen and nitrate (NO_3^-) or other N oxides. Various studies have found high denitrification activity is created by decomposing organic matter which generates anaerobic microsites. This phenomenon may explain some of the high spatial variability of soil denitrification commonly observed.

Soil moisture and temperature. Soil moisture and temperature control soil processes at all levels by governing organic matter decomposition rates, denitrification and nitrification. Some studies have shown that emissions of NO increase with increasing soil temperatures. However, the relationship between temperature and NO fluxes is subject to considerable uncertainty with numerous exceptions in temperate and tropical systems, where no clear relationship was found between temperature and NO fluxes. During denitrification, the ratio of N_2O/N_2 generally increases with decreasing temperatures.

In systems where soil N temporarily accumulates due to wet-dry or freeze-thaw cycles, then early spring, winter and autumn in temperate climates may account for an important part of the annual N_2O emission from agricultural or any other land. Soil-water content influences N_2O and NO_x emissions from all types of soil. In general, microbial activity peaks at 30-60 percent water-filled pore space. Nitrification and associated N_2O and NO_x production also show maximum activity at 30-60 percent water-filled pore space, while optimum conditions for denitrification may occur at 50-80 to 60-90 percent water-filled pore space.

The oxygen and moisture status and gas diffusion in agricultural soils depend on soil texture and drainage. Fine-textured soils have more capillary pores within aggregates than do sandy soils, and so hold soil water more tightly. As a result, anaerobic conditions may be more easily reached and maintained for longer periods within aggregates in fine-textured soils than in coarse-textured soils.

Wetting. The wetting of dry soils causes pulses in N mineralization, nitrification and NO and N_2O fluxes. The alternating drying and wetting of soils enhances the release of N_2O and NO from the soil to the atmosphere, although peaks in N_2O production may decline with subsequent wetting events.

Soil pH. Soil pH has a marked effect on the products of denitrification. Denitrification rates are slower under acid than under slightly alkaline conditions, but the N_2O fraction may be larger at low soil pH, particularly with an adequate nitrate supply. This is commonly attributed to the sensitivity of N_2O reductase to proton activity. Emissions of N_2O and NO decrease with increasing pH in acid soils, and increase when the pH of alkaline soils decreases. Soil pH is a major factor in determining the mechanism of NO formation. In one study, in an alkaline loamy clay soil (pH 7.8) nitrification was the main source of NO, whereas in an acid, sandy clay loam (pH 4.7) denitrification dominated the NO production.

Gas diffusion. Before escaping from the soil to the atmosphere, the N gases diffuse through the soil pore system, where denitrifiers may consume NO and N_2O or plants may take up NO. In situations where high soil water content, impeded drainage, shallow groundwater, soil structure, soil compaction, fine soil texture or soil surface sealing limit gas diffusion, denitrification activity is high. However, such conditions greatly enhance the probability of denitrifiers re-consuming N_2O and NO. Soils close to saturation show high denitrification activity but low N_2O and NO emission. Under such conditions, the oxygen supply is low and N_2O and NO diffusion is limited; in the absence of oxygen as an electron acceptor, denitrifiers consume more N_2O and NO than under more aerobic conditions. Under wet conditions, N_2O uptake from the atmosphere may occur.

Agricultural Management Factors

Crop type. Nitrous oxide fluxes from flooded rice systems during the growing season are generally lower than those from rainfed fields. The cause of this may be the anaerobic conditions prevailing in wetland rice systems. However, aerobic conditions generally prevail during the fallow period after the rice crop, allowing nitrification of NH_4^+ mineralized from soil organic matter and residues of crops and aquatic biota. Observed N_2O emission from drained soils during the post-harvest fallow period is much higher than during the crop season.

Biological N fixation by leguminous crops, such as alfalfa, soybeans, pulses, and clovers, provides an important input of N in many agricultural systems. Although, these crops generally receive no or only small N fertilizer inputs as a starter, studies have demonstrated that leguminous crops show emissions of N_2O that are of the same level as those of fertilized non-leguminous crops.

Fertilizer management. The type of fertilizer influences the magnitude of nitrification and denitrification. Furthermore, the type of fertilizer may influence the magnitude of NO and fluxes. For example, anhydrous ammonia (AA), which is generally injected, commonly shows higher losses than other fertilizers.

Total N gas production and in particular emission may increase with the N application rate. However, denitrification and production may be more closely related to the amount of unused N than to the total amount. Therefore, the timing of fertilizer application is an important factor. Any prolongation of the period when NH_4-based fertilizers can undergo nitrification or NO_3-based fertilizers undergo denitrification, without competition from plant uptake, is likely to increase emissions of NO and N_2O.

Under tropical conditions, the potential for high and NO emission rates exists, but management plays an important role in determining the magnitude of the fluxes. Where fertilizer application is excessive, as in some cash crops such as banana, the N availability for nitrification and denitrification is high, and so the O_2 supply becomes the primary control of NO and N_2O emissions.

Fertilizer application mode and timing influence NH_3

volatilization and the efficiency of plant uptake, and hence the availability of N for nitrification and denitrification. Generally, emissions from subsurface applied or injected N fertilizers are higher than from broadcast mineral fertilizers and animal manure. Compared to broadcasting, subsurface applied N has resulted in higher N_2O but lower NO losses.

Soil and crop management. Inputs from crop residues in agricultural fields are important sources of C and N for nitrification and denitrification. In addition, the action of incorporation may also stimulate mineralization of soil organic matter. Some studies have observed higher denitrification activity and N_2O fluxes after retaining residues on the land compared to removing them. This effect may be related to the effect of residues on the soil surface on moisture conditions of the topmost soil layer. The incorporation of residues also causes accelerated NO emissions.

The effects of tillage on mineralization of soil organic matter are well established. Tillage may also affect the conditions for N_2O and NO emissions from soils. Some studies have observed higher N_2O losses for no-tillage systems compared to conventional tillage. This may be related to higher denitrification activity. However, N_2O losses from no-tillage systems may be lower than from tilled soils where fields have remained uncultivated for a number of years. Some studies report an increase in NO emission caused by ploughing in a temperate climate and under tropical conditions.

Factors Regulating Ammonia Loss

Ammonia is formed constantly in soils because of the biological degradation of organic compounds and NH_4^+ yielding mineral and organic fertilizers. As it is a gas, any NH_3 present in soils, water or fertilizers can volatilize to the atmosphere. However, NH_3 reacts with protons, metals and acidic compounds to form ions, compounds or complexes of varying stability. Ammonia has a strong affinity for water, and its reactions in water are fundamental to regulating the rate of loss. After its application to the soil, the NH_4^+ can remain on the exchange sites, nitrify to

NO_3^-, or decompose to NH_3, depending on soil and environmental conditions:

$$NH^+_{4(adsorbed)} \longrightarrow NH^+_{4(solution)} \longrightarrow NH_{3(solution)} \longrightarrow NH_{3(soil)} \longrightarrow NH_{3(atmosphere)}$$

$$NH_4^+ \longleftrightarrow NH_3 + H^+$$

$$NH_4^+ + 1.5O_2 \longrightarrow 2H + H_2O + NO_2$$

$$NO_2^- + 0.5O_2 \longrightarrow NO_3^-$$

In fertilized fields, the input of NH_4^+ depends on fertilizer type, the rate and mode of fertilizer application, soil moisture content, infiltration rate, CEC and urease activity (in the case of urea). The difference in NH_3 partial pressure between the ambient atmosphere and that in equilibrium with moist soil, floodwater, or the intercellular air space of plant leaves drives ammonia volatilization. The partial pressure of NH_3 in the soil is controlled by the rate of removal of ammonium or NH_3 in solution, or by displacing any of the equilibria in some other way. Wind speed (regulating the exchange between soil and air), temperature, and the pH of the soil solution or irrigation water are important regulating factors, as all three variables affect the partial pressure of NH_3.

Increasing temperature increases the relative proportion of NH_3 to NH_4^+ present at a given pH, decreases the solubility of NH_3 in water, and increases the diffusion of NH_3 away from the air-water or air-soil interface.

In particular, the pH affects the equilibrium between NH_4^+ and NH_3 so that the relative concentration of NH_3 increases from 0.1 percent to 1, 10 and 50 percent as the pH increases from 6 to ~7, ~8 and ~9, respectively. The volatilization process itself produces acidity. The nitrification process can reduce NH_3 volatilization in two ways. Firstly, by decreasing NH_4^+ availability, and secondly by producing acidity.

Other variables influencing NH_3 volatilization include the pH buffer capacity and the CEC of the soil. The CEC is important as the negatively charged CEC absorbs the positively charged NH_4^+. A major part of the soil's NH_3 holding capacity is attributable to soil organic matter.

Other factors include the level of urease activity (in the case of urea application, or urine), the availability of moisture, soil texture, the nitrification rate, and the presence of plants or plant residues. The presence of absorbed or added Ca and certain phosphate compounds may influence NH_3 losses in upland systems. The loss of NH_3 may be lower in the presence of Ca due to micro-environmental pH reduction, $CaCO_3$ formation or Ca-urea double salt formation. Certain acidic phosphate compounds, such as H_3PO_4, $NH_4H_2PO_4$ and $Ca(H_2PO_4)_2$ may precipitate the micro-environmental Ca and change the environment to favour increased NH_3. Both the pH and form of the phosphate added can influence reactions with Ca, and therefore NH_3 losses.

In wetland rice systems, factors such as fertilizer type, rate, time and method of application, floodwater depth, and algal growth exert their influences through the primary variables (ammoniacal N concentration, the pH and temperature of floodwater, and wind speed).

The rate of NH_3 loss depends on the equilibrium vapour pressure of NH_3 in floodwater and on wind speed. Wind speed markedly affects NH_3 volatilization from floodwater. The vapour pressure of NH_3 in floodwater is a function of the ammoniacal N concentration, pH and temperature. The content of aqueous NH_3 in floodwater increases by about a factor of 10 per unit increase in pH (pH range 7.5 - 9.0). Furthermore, water evaporation losses, which may be high under tropical conditions, will generally lead to increasing activity of NH_4^+, thus influencing the chemical equilibria.

Processes that potentially affect the ammoniacal concentration of floodwater and indirectly the NH_3 loss include urease activity, cation exchange and N immobilization. Furthermore, assimilation of NH_4^+ by algae, weeds and rice plants can decrease the quantity of ammoniacal N. The competitiveness of rice plants for NH_4^+ varies with growth stage.

Fertilizer management, through its influence on the concentration of ammoniacal N in the floodwater, has a pronounced effect on overall NH_3 loss. NH_3 losses are generally highest with applications of urea or ammonium sulphate fertilizer 2-3 weeks after transplanting the rice. Volatilization rates are much lower for ammonium sulphate and urea applications a few days before panicle initiation of the rice crop (commonly ~50-60 days after sowing), or at booting (~65-70 days after sowing). The incorporation of urea to puddled soil, when done without standing floodwater, also reduces NH_3 losses. The cause of this reduction may be a better contact of the fertilizer with the soil and an associated increase in the reaction of the ammonium ions formed with the cation exchange sites and the immobilizing microorganisms. Another possible cause may be a reduction of algal growth as a result of lower ammoniacal concentration of the floodwater. The incorporation of urea in the presence of floodwater is effective only in soils with a high CEC.

EMISSION OF CARBON DIOXIDE FROM SOIL

Carbon dioxide (CO_2) is an important greenhouse gas accounting for 60% of the total greenhouse effect. Soil is a major source for atmospheric CO_2. Soil contributes 20% of the total emission of CO_2 to the atmosphere through soil respiration. Besides disturbing the earth's heat budget, emission of CO_2 from the soil results in diminution of soil organic C pool, soil fertility and productivity.

Carbon dioxide is released from the soil through soil respiration, which includes three biological processes, namely microbial respiration, root respiration and faunal respiration primarily at the soil surface or within a thin upper layer where the bulk of plant residue is concentrated, and one non-biological process, i.e. chemical oxidation which could be pronounced at higher temperatures. Processes affecting dynamics of soil carbon are presented in Figure 12.1. Soil microflora contributes 99% of the CO_2 arising as a result of decomposition of organic matter, while the contribution of soil fauna is much less. Root respiration, however, contributes 50% of the total soil respiration. Several studies have shown that factors such as soil texture, temperature, moisture, pH, available C (labile and non-labile components of soil organic matter), and N content of the soil influence CO_2

production and emission from the soil. For root respiration, the source of C is photosynthates and its translocation to the root; while litter fall, root mortality, application of manures and crop residues provide carbon for microbial respiration in the soil. Soil organic matter includes a wide variety of organic substances ranging from freshly added leaves or manures to substances at varying stages of decomposition.

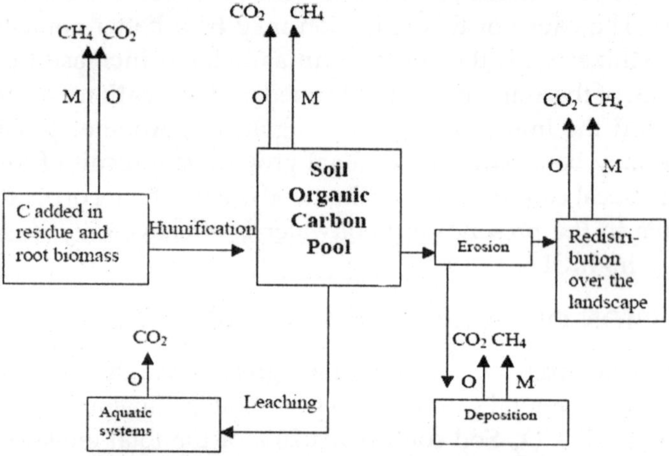

Figure 12.1. Processes affecting soil carbon dynamics.

Changes in agricultural practices can result in changes in both the pool size and turnover rates of soil organic matter. The decomposable and the biomass pools of carbon constitute labile carbon, which declines faster and is restored faster than the non-labile carbon and, therefore, is a more sensitive indicator of carbon dynamics of the system.

Factors Affecting CO_2 Emission from Soil

Temperature. Temperature has a marked effect on CO_2 evolution from the soil. There is a strong relationship between CO_2 evolution and mean daily litter temperature. Increase in CO_2 emission with temperature is a matter of concern, as the possible global warming would increase CO_2 evolution from the soil that would accelerate the depletion of soil carbon and soil fertility. It is estimated that 1°C increase in temperature could lead to a loss of 10% of soil organic carbon in the regions of the world with annual mean temperature of 5°C. While in the regions having a mean temperature of 30°C, 1°C increase in temperature would lead to 3% loss of soil organic carbon.

Moisture. Soil moisture affects soil respiration and hence CO_2 evolution. In general, increasing soil moisture would increase CO_2 evolution up to an optimum level, above which it would reduce CO_2 evolution. Periodic drying and wetting of soil has a pronounced influence on CO_2 evolution. When the soil is rewetted the activity of the microbes, which were in a latent state in the dry soil, increases accompanied by release of air trapped in the soil pores contributing to an increase in CO_2 evolution. Drought condition reduces the soil respiration, while rewetting increases it. Drainage has been shown to increase annual CO_2 emission from forests soils. Apart from the direct effects of temperature and moisture, the interaction of these two assumes great significance in view of the global warming and likely disturbance in precipitation pattern. It is observed that under dry soil condition, soil respiration is greater during the day than at night, while day and night soil respiration rates are similar when the soil is wet. This is attributed to a reduction in the variability in soil temperature when the soil is wet.

Diurnal, seasonal and spatial variability. There are considerable variations in CO_2 emission during different periods during the day and night. Soil respiration values at night always higher than the day values due to the presence of higher relative humidity at night, which favours the activity of microbiota, and to high soil temperature at the beginning of night. Soil temperature, which accounts for most of the temporal variability of CO_2 efflux, is by far the most influential factor controlling soil respiration rate. Soil respiration is maximum during the growing season coinciding with the period of maximum growth of crops. Seasonal CO_2 flux is maximum in spring followed by summer, autumn and winter. In spring, neither temperature nor moisture is limiting, resulting in better crop growth and higher soil respiration. In summer, moisture becomes limiting and in winter the limiting factor is temperature, hindering crop growth and soil respiration. On comparing summer and winter

CO_2 fluxes, CO_2 evolution is generally higher in summer than in winter, while in autumn it is intermediate. Spatial variability of soil respiration may usually occur at a scale smaller than 15 cm, which is attributed to the contribution of plant roots as maximum soil respiration during the growing season coincided with the period of maximum root growth.

Soil texture. Soil texture affects the spread of microbial propagules and the growth of bacteria and fungi through the supply of air and moisture, and thus affects formation of CO_2. Water infiltration and gas diffusion rates are also greatly influenced by soil texture and thereby CO_2 formation and emission.

Soil pH. Hydrogen ion activity (pH) of soil has a marked effect on the growth and proliferation of soil microbes. Low pH on soil has adverse effect on microbial activity which lower the respiration rate and consequently lower CO_2 evolution.

Salinity. Excess amounts of salt have adverse effects on physical, chemical and microbiological processes in soil, including C and N mineralization and enzyme activities, which are crucial for decomposition of organic matter. Organic manure amendment, however, increase the biologically evolved CO_2 from these soils, except at a very high salinity where the emission remained low.

Atmospheric pressure. Low atmospheric pressure increases the CO_2 emission. Decrease in atmospheric pressure trigger the escape of CO_2 stored in the soil profile to the atmosphere.

Organic manure application. Application of organic manure in soil can increase CO_2 emission. Soluble organic C in the soil is an immediate source of C for soil microorganisms, which in turn emit CO_2. Large quantities of organic manure that are added to agricultural soils every year for supplying nutrients to crops may contribute significantly to CO_2 emission. Application of sewage sludge can also enhance CO_2 emission. The application of straw on the soil surface increases CO_2 fluxes.

Fertilizer application. Application of nitrogenous fertilizer affects CO_2 emission, (i) directly by providing nitrogen to crops and microbes, and (ii) indirectly by influencing soil pH, which influences microbial activity. There may be about 30–40% reduction in CO_2 emission with the addition of NH_4NO_3 fertilizer, which is due to reduced microbial respiration by increased acidity. Increase in the supply of NH_4-N leads to reduction in the decay of organic matter and loss of C. The addition of nitrogenous fertilizer leads to a decrease in CO_2 evolution.

Use of nitrification inhibitors. Nitrification inhibitors, viz. dicyandiamide, nitrapyrin, thiosulphate and acetylene inhibit nitrification, a microbially mediated process. Therefore, they may act on microbes engaged in the oxidation of organic carbon. In a greenhouse study, encapsulated calcium carbide (a slow release source of acetylene) reduced CO_2 emission and appeared to be effective in minimizing emission of this greenhouse gas from flooded rice. However, studies on the effect of nitrification inhibitors on CO_2 emission are limited and need further investigation.

Crops. Presence of crops influences CO_2 production and emission from the soil. Production of CO_2 is approximately 2 to 3-fold greater in cropped soil compared to bare soil. Within different crops also there is variability in CO_2 production.

Tillage. More CO_2 emission can occur from a tilled than from an undisturbed soil (no tillage) as tillage produces a soil microenvironment favourable for accelerated microbial decomposition of plant and animal residues. Tillage breaks down soil aggregates, helps in mixing soil and organic particles, improves infiltration and water-holding capacity and thereby increases CO_2 production, while reduced tillage is reported to reduce emission of CO_2 due to less ploughing of soil and keeping the soil organic C unexposed. Low or zero CO_2 fluxes under no-tillage are associated with reduced gas diffusivity and air-filled porosity, while increased CO_2 emission with ploughing is due to degassing of soil CO_2. Tillage enhances C mineralization and atmospheric fluxes, suggesting that tillage intensity should be decreased to reduce C loss from the soil.

CO_2 Emission from Different Ecosystems

It may be noted that an ecosystem may have low rate of CO_2 emission but a large area extent, and therefore, total CO_2 flux could be more. Effluxes of CO_2 from different biomes vary markedly with the taiga biome having the highest amount of CO_2 emission and the polar desert, the least.

Peat lands. Peat lands and other wetlands play a major role in the regulation of the atmospheric CO_2. Peat lands of northern latitudes represent an enormous store of organic C and are primarily sinks of CO_2, as assimilation of carbon through photosynthesis generally exceeds the release through the decomposition of plant litter. Efflux of CO_2 from peat land to the atmosphere is a function of plant root respiration and decomposition of plant material and peat in the peat profile. The rate of carbon dioxide production is greatest in the upper part of the peat profile and is related to the botanical origin of the peat and also to temperature, water content and availability of O_2. On the contrary, large boreal peat land ecosystems sequester C from the atmosphere due to a low oxygen pressure in waterlogged peat. However, when such soils are brought under cultivation, they become large emitters of CO_2. In response to global climate change resulting in elevated temperatures and lower water table, peat lands may become a net source rather than a sink of CO_2.

Forests. Forest is also a major sink of atmospheric CO_2.

Deserts. There is substantial emission of CO_2 from desert soil because of the presence of calcium carbonate. The two major factors influencing CO_2 evolution from desert are moisture and temperature. Moisture not only limits microbial activity in the soil but also serves as a constraint for plant growth and hence, input of C to the soil.

Sequestration of C in Soil

Soil management strategies for C sequestration include three approaches. First, management of soil to maintain higher than existing levels of soil organic matter through reduced tillage and no tillage practices. Secondly, to manage carbon-degraded soils so as to restore soil organic matter levels. Wastelands soils have relatively high potential for accumulating organic carbon in vegetation and in soil, if suitable trees are grown along with proper soil conservation measures. Promoting agro-forestry in these lands is highly desirable. Thirdly, enlarging soil organic matter pools by improving soil fertility. Soil processes could be managed so that litter production exceeds decomposition. Such an approach of increasing carrying capacity of soils is difficult as it is determined by factors such as climate, that cannot be managed.

Another approach could be to increase the passive or inert fraction of soil organic pool. This implies taking soil organic carbon out of the humification–mineralization cycle. This can be achieved through the increase of sub-soil organic carbon and microaggregation. Sub-soil organic carbon can be increased by growing deep-rooted plants and deep ploughing, while micro-aggregation can be increased by using soil conditioners, long-chain polymers and earthworms. Recently, eco-friendly farming practices like organic farming and precision farming are getting popular worldwide. Organic farming, where all the NPK requirement of crops is supplied through organic sources, has a great potential to enrich soil with organic carbon through sequestering C in soil. In precision farming, site-specific crop management is done, keeping in account the variability in soil and micro-climatic conditions in the field, and wasteful use of inputs is avoided.

METHANE EMISSIONS FROM RICE ECOSYSTEM

The most important source of methane (CH_4) emission is the mineralization of organic matter through microbial agencies under anaerobic conditions. These are generally encountered in lowland soils undergoing submergence as in rice culture and in marshes. Soil under submergence are a net source of methane with seasonal variations in the emissions rates less than 0.1 to 0.30 g/m^2 per day. This could be considerably higher under tropical conditions due to more intense microbial activity taking place as a result of elevated temperatures. However, under drought conditions these microbes observed a reverse process in which soils consumed atmospheric methane at rates between 0.1 to 0.5 g/m^2 per day. This illustrates the complexity of the process which regulates the net flux of methane in the conditions encountered under upland rice cultivation, where soils are undergoing alternate drying and submergence.

CH_4 mainly produced in the reduced topsoil where easily degradable organic substances are available. If the subsoil is saturated with water, B horizon may also become a source of CH_4. Methanogenic bacteria are strict anaerobes and intolerant of O_2 exposure for growth, although they may not be killed by short exposure to oxygen.

About twenty genera of methane-producing bacteria have been described but only a few, including *Methanobacterium* and *Methanosarcina* have been isolated from rice soils. *Mathenospirillium* and *Methanocorpusculum,* which were isolated from freshwater sediments, as well as methanogens found in endosymbionts in sapropelic amoebae may also be present in wetland soils. Most described species are hydrogenotrophs and use formate or acetate while methylotrophs use compounds such as methanol or methylated amines. Most of them can fix CO_2 and need H_2 as their electron donor (energy source) even if they use acetate as a carbon source.

In wetland rice soils, the major substrates for CH_4 production are acetate and H_2-CO_2. Conditions for high CH_4 production in wetland soils depend on into six parameters i.e. water regime, Eh/pH buffer, carbon supply, temperature, texture and mineralogy and salinity. Rice soils that are prone to CH_4 production mainly belongs to the orders of Entisols, Inceptisols, Alfisols, Vertisols and Mollisols.

Reductions in rice agriculture remain largely untenable as on methane emission reduction strategy. However, through a more integrated approach to rice paddy irrigation and fertilizer application substantial reductions remain possible. Many rice varieties can be grown under much drier conditions than those traditionally employed, with big reductions on methane emission without any loss in yield. Additionally, there is the great potential for improved varieties of rice, able to produce a much larger crop per area of rice paddy and so allow for a cut in the area of rice paddies, without a cut in rice production. Finally, the addition of compounds such as ammonium sulphate, which favour activity of other microbial groups over that of the methanogens, has proved successful under some conditions.

SULPHUR GAS EMISSIONS FROM WETLANDS

Wetlands emit a range of sulphur gases and they vary according to factors such as salinity, wetting-drying regime, soil type and diurnal cycles. Three main types of S gases can be emitted by wetlands are- hydrogen sulphide (H_2S), volatile organic sulphur compounds (VOSC) and sulphur dioxide (SO_2). These differ in the way they are produced.

Hydrogen sulphide

Hydrogen sulphide is produced by microorganisms in the water columns and sediments of wetlands by the process of sulphate reduction. Essentially, sulphate reducing microorganisms can use SO_4^{2-} instead of oxygen to respire, producing H_2S as an end-product. Most wetland sediments will be anoxic (without oxygen) at depth because bacteria use the oxygen to respire when feeding on (i.e., decomposing) the organic matter that accumulates on the bottom of the wetland. This means there will be at least some sulphate reduction occurring. While a number of factors control the rates of sulphate reduction, rates tend to be higher in saline wetlands because more sulphate is available. Another mechanism to produce H_2S is the decomposition of organic matter rich in organic S. H_2S is a toxic gas and can cause illnesses from chronic exposure at low concentrations (irritations of airways, etc.) and be lethal at higher concentrations.

Volatile Organic Sulphur Compounds (VOSC)

A wide variety of volatile organic sulphur gases occur in the environment. VOSCs can be produced by a wide variety of mechanisms including decomposition of organic matter, releases by algae, salt marsh environments, the degradation of dimethylsulphoniopropionate (DMSP) an organic osmolyte (a substance that is part of an organism's salt balance), and as detoxification by-products when organisms are exposed to H_2S. There is some evidence that naturally saline wetlands and marshes tend to emit more VOSCs than their freshwater counterparts, but the reasons for this are not clear.

Sulphur dioxide

When previously anoxic sediments are exposed to oxygen, sulphur dioxide (SO_2) can be produced during the oxidation of sulphides (e.g., pyrite) by oxygen. This process can also be catalysed by another group of microorganisms, the 'sulphide oxidisers'. Large quantities of SO_2 are also emitted by fossil fuel burning and it is a key ingredient for the formation of smog in cities.

KEY BIOGEOCHEMICAL PROCESSES

Inorganic S Cycle (under wet conditions)

There are two key features of the sulphur cycle under wet conditions:

1) the accumulation of sulphides and organic-S in the sediments, and
2) the trapping of S gases emitted from the sediments by the water column.

In most wetlands, the decomposition of organic matter in the sediments consumes much oxygen and anoxic conditions prevail within a few millimetres below the sediment-water interface. Under such anoxic conditions, dissimilatory sulphate reduction occurs and H_2S is produced as an end-product. Much of the H_2S produced in the sediments will react with Fe and other metals to form sulphide minerals, but some H_2S can also escape to the water column by diffusion from the sediments. However, most of the H_2S diffusing from sediments to the water column will rapidly react with O_2 (eventually producing sulphate) and thus not reach the atmosphere.

A few other factors could influence whether or not H_2S will be emitted from wetlands, including the diurnal wetland O_2 cycle, density stratification, pH and ebullition (bubble formation).

Diurnal O_2 cycle. It is common for small wetlands to have large fluctuations in O_2 concentration over a daily cycle. Oxygen is added to the water column of wetlands by diffusion from the atmosphere and by photosynthesis by plants, algae and some bacteria. In return, O_2 is consumed in wetlands by the decomposition of organic matter and aerobic respiration. Because photosynthesis does not occur at night but respiration and decomposition do, there is a tendency for small wetlands to have lower O_2 concentrations at night, especially in summer when warm temperatures increase the rates of decomposition and respiration. It can be hypothesised that H_2S produced in sediments will be more likely to escape to the atmosphere at night when low O_2 concentration in the water column would decrease water column H_2S oxidation rates.

Density stratification. Many wetlands have periods when the water column is stratified into layers with different densities because of differences in temperature and salinity. When stratified, there is a tendency for bottom waters to become anoxic and accumulate H_2S because O_2 consumption rates in sediments can be rapid relative to the rate of diffusion of O_2 from the overlaying water layer. In shallow wetlands, such stratification is seldom long lasting when it occurs because wind can efficiently mix the water column. However, when wetlands destratify rapidly, it is possible that a pulse of H_2S from bottom waters could be rapidly released to the atmosphere.

pH. H_2S is a "weak acid" and a proportion present in the water column will tend to dissociate into either HS^- or S_2^- at different pH. In general, at neutral and alkaline pH, the preferred forms in water will be HS^- and S_2^- (which are not volatile) while at more acidic pH, H_2S becomes favoured. Thus, it should be expected that a higher proportion of the H_2S produced in wetland would escape to the atmosphere when wetlands are acidic.

Ebullition. Diffusion is not the only process that can transfer H_2S produced in sediments to the water column. When water becomes supersaturated with a given gas (or set of gases), bubbles will tend to form. Bubble formation in unconsolidated sediments at the sediment–water interface is especially common in summer, when decomposition rates (and the production of CO_2, N_2, methane and other gases) are higher and the solubility of gases in the warmer water is lower. The formation and release of bubbles in sediments could be an effective mechanism to emit H_2S produced in sediments to the atmosphere in some wetlands.

Organic S Cycle (under wet conditions)

The organic sulphur cycle in wetlands is not as well known as its inorganic counterpart. Sulphur can be integrated in organic matter by the process of assimilatory SO_4^{2-} reduction. This occurs when algae and bacteria consume sulphate and convert it to protein-sulphur within their cells. Organic sulphur accumulates on the bottom wetlands when plants and animals die. Because decomposition rates are slower under anoxic conditions, there is tendency for organic matter (and organic-S) to slowly accumulate in wetland sediments.

VOCS will be produced in wetland water and sediments when wet, particularly by releases from algae and decomposition in the sediments. It is possible that some of the VOSCs produced will be consumed by bacteria in the water column. How much of the remaining VOSCs will be released to the atmosphere is unknown. What factors control the rates of production and emissions for different compounds is not well known. However, in some studies, VOSC emission rates tend to be higher in saline wetlands. In addition, VOCS emission rates are related to other factors such as temperature, diurnal cycles, degree of sediment wetness and sediment texture.

Change in S Cycling (when wetlands are drying)

Two key factors could influence the S gas emission rates when wetlands are drying-

1) the exposure of previously anoxic sediments to oxygen, and
2) the loss of the overlying water column as a "trap" for S gases emitted from the sediments.

The proposed consequences for the changes in these two environmental factors on the inorganic and organic S cycle are outlined below.

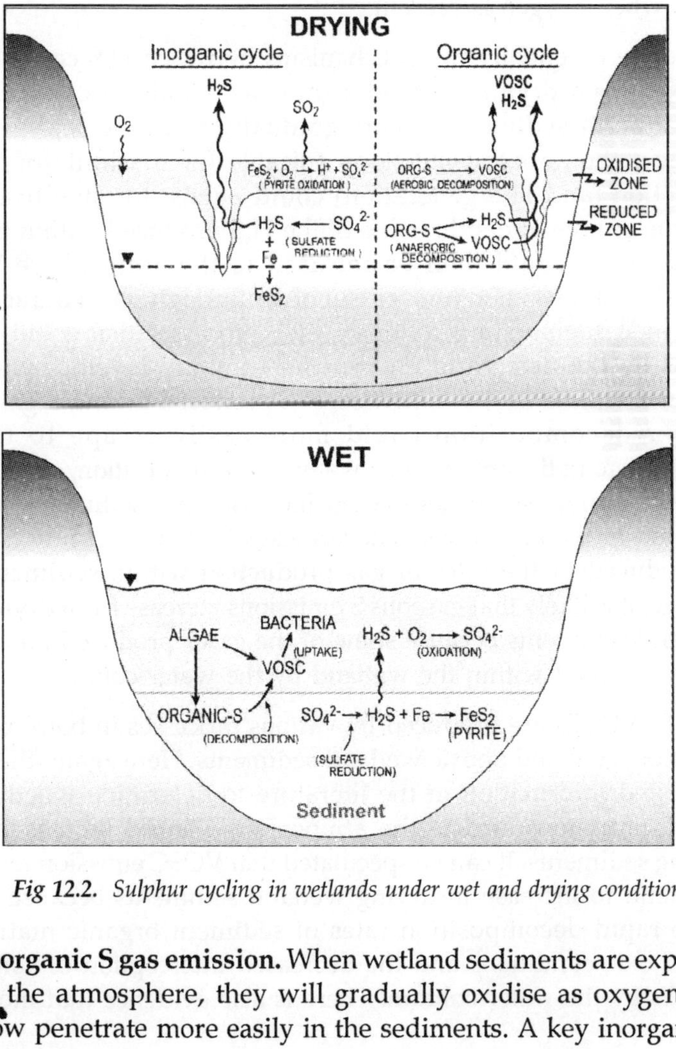

Fig 12.2. Sulphur cycling in wetlands under wet and drying conditions

Inorganic S gas emission. When wetland sediments are exposed to the atmosphere, they will gradually oxidise as oxygen can now penetrate more easily in the sediments. A key inorganic S process under these conditions will be the tendency to oxidise sulphides stored in sediments, producing both sulphuric acid and sulphur dioxide. The rates at which sulphide oxidation proceed is dependent on the soil texture (lower in clay compared to sandy sediments) and the frequency of desiccation features (which increase the rates of oxygen diffusion in the sediments), residual moisture, and ambient pH. Thus, SO_2 as opposed to H_2S is the main S gas emitted by the inorganic S cycle when oxygen is present.

There are potentially two mechanisms by which H_2S could be emitted from drying wetland sediments. Firstly, because the oxidation of sediments occurs gradually from the sediment surface, the anoxic conditions suitable for dissimilatory S reduction (and H_2S generation) could persist for significant lengths of time at depth. Some of the H_2S produced within the anoxic zone could then be released to the atmosphere by diffusion or advection (movement of air through pores induced by winds, temperature contrasts, etc), processes that could be aided by the desiccation features that often develop when wetland sediments dry. Likewise, H_2S produced during organic matter decomposition could more easily escape to the atmosphere in the absence of a water cover. In addition, organic matter decomposition rates could increase once sediments are exposed to the atmosphere. Regardless of whether H_2S or SO_2 is produced or the rates of gas production within sediments change, it is likely that gaseous S emissions increase from drying wetland sediments because some of the gases produced are no longer trapped within the wetland by the water column.

VOSC. VOSCs are produced by various processes in both oxic (with oxygen) and anoxic wetland sediments. Here again, there is limited information in the literature to determine whether VOSC emission rates to the atmosphere should increase in drying sediments. It can be speculated that VOSC emission rates will tend to increase in drying wetland sediments because of more rapid decomposition rates of sediment organic matter under oxic conditions and the decreased interception of some VOSCs by the water column. There is also evidence that more

VOSC's are produced in sulphide rich environments. This could occur, for example, when microorganisms produce VOSC's as a detoxification mechanism against H_2S or by chemical reactions between H_2S and organic matter.

MANAGEMENT OF THE GASEOUS EMISSIONS

Emission of greenhouse gases has become a matter of great concern because of the future projections of global warming and related effects on biological life. Soil is a major sink for many gaseous pollutants. The removal by soil of SO_2, NO_2, O_3, H_2S, methyl mercaptan and NH_3 appears to be primarily by abiotic processes but once gases come into contact with soil solution, the soluble products can be utilized by microbiota. Microorganisms are responsible for removal of N_2O, CO, C_2H_4 and C_2H_2, dimethyle sulphide, carbonyl sulphide and carbon disulphide and a minor sink for SO_2 which can be removed by some fungi. Removal of CO occurs by fungal and bacterial co-metabolic oxidation of CO_2. Main sink for methane in atmosphere is troposphere, where it is chemically destroyed. Soil is not a significant sink for methane.

SUMMARY

In the last few decades there has been an increase in the emission of naturally occurring, radiatively active gases like nitrous oxide (N_2O), carbon dioxide (CO_2) and methane (CH_4) popularly known as 'greenhouse gases'. These gases trap the outgoing infrared radiation from the earth's surface. This process, generally referred to as the greenhouse effect, adds to the net energy input of the lower atmosphere, and leads to regional and global changes in climatic parameters like temperature and rainfall. Human health, terrestrial and aquatic ecological systems, agriculture, forestry, fisheries, and water resources are sensitive to changes in climate.

The major factors regulating NO emission are N application rate, fertilizer type, soil organic C content and soil drainage. Management factors that influence N_2O emission are crop type (with N_2O emission increasing in the order- grass, non-leguminous annual crops, leguminous crops) and type of fertilizer (with the highest values being for anhydrous ammonia and mixtures of mineral and organic fertilizers). Ammonia is formed constantly in soils because of the biological degradation of organic compounds and NH_4^+ yielding mineral and organic fertilizers.

The factors influencing CO_2 production and emission from the soil include inherent properties of the soil like texture, moisture, pH and salinity, soil microbial activity and root respiration. Besides these, external factors (seasonal effect and atmospheric pressure) and manipulation of soil environmental conditions, viz. tillage, irrigation, fertilizer and manure application also have an effect on CO_2 production and emission. Soil management practices like increasing soil organic carbon content, reduced tillage, manuring, residue incorporation, improving soil biodiversity, micro-aggregation and mulching play an important role in sequestering C in soil.

Rice ecosystem is a major source of CH_4. Global CH_4 emission estimates from rice ecosystem range from 29 to 61 Tg/yr. It has been estimated that global rice production must almost double by the year 2020 in order to meet the growing demand and this may increase CH_4 fluxes by up to 50%. Many rice varieties can be grown under much drier conditions than those traditionally employed, with big reductions on methane emission without any loss in yield. There is a need for careful evaluation of the source strength of this ecosystem, and of the influence of soil, water and crop management practices on both grain yield and green house gas fluxes.

Environmental factors controlling the rates of emission of S-gases include sediment texture, organic matter content, sulphide content, water content, salinity, pH, time of day, temperature and the presence or absence of a water column. There is very limited study on environmental controls on S gas emissions. Therefore, it is necessary to determine what inorganic and organic sulphur gases are emitted from wetlands and their relative significance of dissimilatory sulphate reduction and organic matter decomposition as sources of H_2S emission during wetland drying events.

Reconciling the goals of ensuring ample food supply, enabling adequate farm income and reducing the adverse environmental impact of increased fertilizer use requires the adoption of sound agricultural practices. The adoption of balanced fertilization would be one of the measures for achieving higher yields without incurring accelerated toxic gaseous losses to the environment.

13

Radionuclides Contamination in Soils

All soil used anywhere in the world for agriculture contains radionuclides to a greater or lesser extent. Typical soils contain approximately 300 kBq/m^3 of ^{40}K to a depth of 20 cm. This radionuclides and others are then taken up by crops and transferred to food, leading to a concentration in food and feed of between 50 and 150 Bq/kg. The ingestion of radionuclides in food is one of the pathways leading to internal retention and contributes to human exposure from natural and man-made sources. Excessive contamination of agricultural land, such as may occur in a severe accident, can lead to unacceptable levels of radionuclides in food.

RADIONUCLIDES

Nuclides of any element that have atomic number greater than bismuth-83 is unstable and therefore radioactive. They are called heavy nucleons. Over sixty radionuclides can be found in nature and they can be placed in three general categories- the primordial, cosmogenic and anthropogenic. Americium-241, ceasium-137, cobalt-60, iodine-129 and -131, plutonium, radium, radon, strontium-90, technetium-99, tritium, thorium and uranium, are the twelve radionuclides that occur naturally. Caesium for example, is an alkali metal like potassium and resemble it metabolically. Whereas, potassium is an essential element for man. Stable caesium-133 is fairly rare in the biosphere and in geological occurrence. The radioactive isotope caesium-137 is produced in nuclear fission and is one of the most significant products. The fission yield is relatively high, about 6 atoms per 100 fissions, independent of the type of fission in uranium or plutonium. It has a radioactive half-life of 30.17 years and its beta decay is accompanied by a gamma ray of moderate energy.

Sources of Radionuclides in Soil

Nuclear fission in connection with atomic weapons testing and nuclear power generation provides some of the sources of soil contamination. To the naturally occurring radionuclides in soil such as 40K, 87Rb and 14C, a number of fission products have been added. However, only two of these are sufficiently long-lived to be of significance in soils: strontium-90 with half life of 29.1 years and caesium-137 with the half life of 30 years. The normal soil levels of the fission radionuclides are not high enough to be hazardous. Even during the peak periods of weapons testing, soils do not contribute significantly to the level of these nuclides in plants. Atmospheric fallout on the vegetation is the primary source of radionuclides in the food chain. Only in the event of a catastrophic supply of fission products toxic soil levels of 90Sr and 137Cs can be expected. Fortunately, considerable research has been accomplished on the behaviour of these two nuclides in the soil- plant system. In addition to radionuclides added to soil as a result of weapons testing and accidents such as that which occurred at Chernobyl (Ukraine), soil may interact with low level radioactive waste materials that have been buried for disposal. Even though the materials may be in solid form when placed in shallow land burial pits, some dissolution and subsequent movement in the soil are possible.

Plutonium, uranium, americium, neptunium, curium and caesium are among the elements whose nuclides occur in radioactive wastes. Nuclides in wastes vary greatly in water solubility, uranium compounds being quite soluble, compounds of plutonium and americium being relatively insoluble and ceasium compounds intermediate in solubility. Ceasium, a positively charged ion, is adsorbed by soil colloids. Uranium is thought to occur as a UO^{2+} ion that is also adsorbed by soil. The charge on plutonium and americium appears to vary, depending on the soil. There is considerable variability in the actual uptake by plants of these nuclides from soils, depending on such properties as pH and organic matter content. Uptake from soils by plants is generally lowest for plutonium, highest for neptunium and intermediate for americium and curium. Fruits and seeds have generally much lower of these nuclides than are leaves, suggesting that grains may be less contaminated by nuclides than forage crops and leafy vegetables.

Reactions of Radionuclides in Soil

Residence times for turnover of radionuclides in a defined soil layer are derived generally on the basis of the total contents of radionuclides rather than on the basis of specific fractions extractable with defined extracting agents. As a result, models for radionuclides movement in soils can generally be simplified as shown in Figure 13.1.

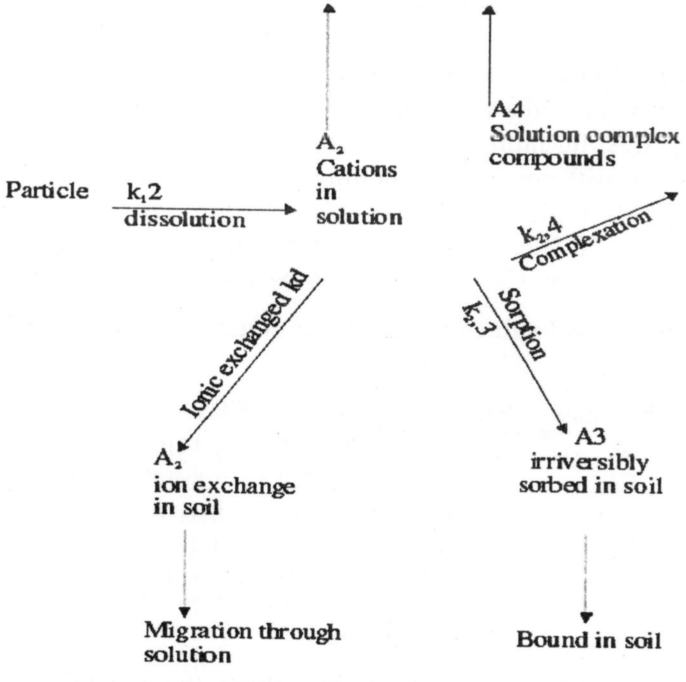

Figure 13.1. Radionuclide transformation processes in soil.

Radionuclides present as original deposited particles irreversibly sorbed to soil particles or present in ion exchange sites on soil particles are subject to processes involved in particle migration. Radionuclides present in soil solution or as organic complexes are subject to processes involved in solution transport.

The solubility of most radionuclides, which are formed in nuclear reactors or other nuclear devices, is low. Consequently, the major part of radionuclides released into the environment will finally accumulate in either the upper layer of soils or interstitial system of sediments in aquatic systems. In general, Pu and Am hardly migrate in soil. Neptunium and strontium migrate over short distance that is about 1 cm in a few years. The migration of Cs depends very much on soil characteristics, usually it is very strongly adsorbed and migrates less than Sr. Migration can be described by chromatography. The soil is divided into a number of layers and radionuclides are assumed to be present in the top layer and partly in solution flow from the upper layer to the second layer. Parts of the radionuclides are adsorbed in the second layer, depending on Kd values (Table 13.1). Despite limitations, Kd is one of the best variables to indicate differences in leaching behaviour in soils of the temperate region. The mass conservation equation, over a period of 20 years is used to indicate the vertical migration velocities for Sr, Cs, Pu and Am to be 0.3 to 0.5 cm per year.

Table 13.1. Kd values for some radionuclides for agricultural soil.

Radionuclides	Sand	Loam	Clay	Organic
Sr	$1.4x10^1$	$2.0x10^1$	$1.1x10^2$	$1.15x10^2$
Cs	$2.7x10^2$	$4.5x10^3$	$1.8x10^3$	$2.7x10^2$
Np	$4.1x10^0$	$2.5x10^1$	$5.5x10^1$	$1.2x10^3$
Pu	$5.5x10^2$	$12x10^3$	$4.9x10^3$	$1.8x10^3$
Am	$2.0x10^3$	$9.9x10^3$	$8.1x10^3$	$1.1x10^5$

Source: Knox et al. (2000), Iskauder (1992).

A very simple approach in describing the availability of radionuclides is the apparent half-life concept. The assumption is that after a certain period, half of the radionuclides disappear from a certain layer. A refinement of this concept is that of the residence time. The soil is assumed to be distributed into different layers. The radionuclides move through subsequent layers, staying in each layer with a certain residence time. The advantage of the approach is that the processes, which cause migration, do not have to be considered. The disadvantage is that diffusion, dispersion and vertical transport cannot be distinguished.

ENVIRONMENTAL CONTAMINATION

The radionuclides contaminants of most significance in agriculture are those which are relatively highly taken up by crops, have high rates of transfer to animal products such as milk and meat, and have relatively long radiological half-lives. However, the ecological pathways leading to crop contamination and the radioecological behaviour of the radionuclides are complex and are affected not only by the physical and chemical properties of the radionuclides but also by factors which include soil type, cropping system (including tillage), climate, season and, where relevant, biological half-life within animals. The major radionuclides of concern in agriculture following a large reactor accident are ^{131}I, ^{137}Cs, ^{134}Cs and ^{90}Sr. Direct deposition on plants is the major source of contamination of agricultural produce in temperate regions. While the caesium isotopes and ^{90}Sr are relatively immobile in soil, uptake of roots is of less importance compared with plant deposition. However, soil type (particularly with regard to clay mineral composition and organic matter content), tillage practice and climate all affect propensity to move to groundwater. The same factors affect availability to plants insofar as they control concentrations in soil solution. In addition, because caesium and strontium are taken up by plants by the same mechanism as potassium and calcium respectively, the extent of their uptake depends on the availability of these elements. Thus, high levels of potassium fertilization can reduce caesium uptake and liming can reduce strontium uptake.

The releases during the Chernobyl accident contaminated about 125000 km^2 of land in Belarus, Ukraine and Russia with radiocaesium levels greater than 37 kBq/m^2, and about 30000 km^2 with radiostrontium greater than 10 kBq/m^2. About 52000 m^2 of this total were in agricultural use; the remainder was forest, water bodies and urban centres. While the migration downwards of caesium in the soil is generally slow, especially in forests and peaty soil, it is extremely variable depending on many factors such as the soil type, pH, rainfall and agricultural tilling.

Soil contamination. The radionuclides are generally confined to particles with a matrix of uranium dioxide, graphite, iron ceramic alloys, silicate-rare earth, and silicate combinations of these materials. The movement of these radionuclides in the soil not only depends on the soil characteristics but also on the chemical breakdown of these complexes by oxidation to release more mobile forms. The bulk of the fission products are distributed between organo-mineral and mineral parts of the soil largely in humic complexes.

Ordinary ploughing disperses the radionuclides more evenly through the soil profile, reducing the activity concentration in the 0-5 cm layer and crop root uptake. However, it does spread the contamination throughout the soil, and the removal and disposal of the uppermost topsoil may well be a viable decontamination strategy.

In Europe, a variation in the downward migration of ^{137}Cs has been seen, from tightly bound for years in the near-surface layer in meadows, to a relatively rapid downward migration in sandy or marshy areas. For example, the greatest deposition in Switzerland and the soil there has fallen to 42% of the initial ^{137}Cs content in the six years after the accident, demonstrating the slow downward movement of caesium in soil. There, the ^{137}Cs from the accident has not penetrated to a depth of more than 10 cm, whereas the contribution from atmospheric nuclear weapon tests has reached 30 cm of depth.

In Belgium, the average body burden of ^{137}Cs measured in adult males increased after May 1986 and reached a peak in late 1987, more than a year after the accident. This reflected the ingestion of contaminated food. The measured ecological half-life was about 13 months. A similar trend was reported in Austria. In short, there is a continuous, if slow, reduction in the level of mainly ^{137}Cs activity in agricultural soil.

Forests contamination. Forests are highly diverse ecosystems whose flora and fauna depend on a complex relationship with each other as well as with climate, soil characteristics and topography. Because of the high filtering characteristics of trees, deposition is often higher in forests than in agricultural areas. When contaminated, the specific ecological pathways in forests often result in enhanced retention of contaminating radionuclides. The high organic content and stability of the forest floor soil increases the soil-to-plant transfer of radionuclides with the result that lichens, mosses and mushrooms often exhibit high concentrations of radionuclides.

The transfer of radionuclides to wild game in this environment could pose an unacceptable exposure for some individuals heavily dependent on game as a food source. This became evident in Scandinavia where reindeer meat had to be controlled. In other areas, mushrooms became severely contaminated with radiocaesium.

Forest initially played the role of a filter in intercepting the fallout. The leaf litter constitutes a relatively closed medium, since the movement of radionuclides towards the subsoil and towards the arboreal vegetation is estimated at less than 2% per year. Deciduous trees have a tendency to accumulate more radioactive material than conifers.

Water contamination. Radioactive contamination of water bodies occurred in the primary forms like the fallout of radioactive aerosols on the surface of water basins, contact of the contaminated air masses with water surface, and as a result of the secondary effects such as radioactivity washout of the surface of water catchement areas, inflow of contaminated water from the more contaminated water bodies and areas into the less contaminated ones, mass exchange between bottom sediment and aquatic masses, discharge of contaminated subsurface water into the surface water bodies, etc.

Chernobyl radionuclides were detected in various rivers of the USSR and Western Europe. Recovery of pre-accidental levels of background radioactive contamination was rather slow in many rivers, whereas it has not occurred in some rivers even 12 years after the fallout. Notwithstanding the current very low levels of contamination in these rivers, the content of ^{137}Ñs and ^{90}Sr is found to increase regularly at rains and spring run-offs.

COUNTER-MEASURES

Mechanical soil cultivation. Counter-measures based on mechanical soil cultivation included removal of soil, ploughing or disking and rotary cultivation of soil upper layer. A major advantage is that, because of their low mobility, the majority of deposited radionuclides can be removed effectively by skimming off a relatively shallow layer of topsoil. Removal of soil upper layer (0-10 cm) can not be applied on soil with thin humus horizon. Ploughing, standard or deep, is used, so that skim and burial ploughing and disking and rotary cultivation of soil upper layer.

Agrochemical measures. It include -

i) liming of acid soils

ii) application of increased doses of K and P-K fertilizers

iii) supplementing soils with natural sorbents (different kinds of clay minerals)

iv) use of organic fertilizers.

The development of agrochemical counter-measures is based on reducing biological availability of radionuclides. Radionuclides in the topsoil are potentially available for uptake by plant roots, although plant uptake of radionuclides is controlled by a variety of factors. The rate of uptake differs substantially for different radionuclides and soil types and depends on the physico-chemical processes in the soil that govern their availability for the plants.

Plant protection counter-measures. Plant protection measures from insect, diseases and weeds in contaminated areas with

^{137}Cs deposition 555-1480 kBq/m^2 are based on the restricted assortment of protection means. The application of plant protection means results in the improvement of the plant physiological processes.

Forest Management. Different strategies have been developed for combating forest contamination. Some of the more effective include restriction of access and the prevention of forest fires. One particularly affected site, known as the "Red Forest". This was a pine forest in which the trees received doses up to 100 Gy, killing them all. An area of about 375 ha was severely contaminated and in 1987 remedial measures were undertaken to reduce the land contamination and prevent the dispersion of radionuclides through forest fires. The top 10-15 cm of soil were removed and dead trees were cut down. This waste was placed in trenches and covered with a layer of sand. A total volume of about 100000 m^3 was buried, reducing the soil contamination by at least a factor of 10. These measures, combined with other fire prevention strategies, have significantly reduced the probability of dispersion of radionuclides by forest fires. The chemical treatment of soil to minimise radionuclide uptake in plants may be a viable option and, as has been seen, the processing of contaminated timber into less contaminated products can be effective, provided that measures are taken to monitor the by-products.

Changes in forest management and use can also be effective in reducing dose. Prohibition or restriction of food collection and control of hunting can protect those who habitually consume large quantities. Dust suppression measures, such as re-forestation and the sowing of grasses, have also been undertaken on a wide scale to prevent the spread of existing soil contamination.

Criteria for Rating Counter-measures

Different criteria should be used for assessing efficiency of counter-measures. First we have the *radioecological criterion*, based on the decrease of concentration in agricultural products and in the food chain. Then the radiological (or dosimetric) criterion is placed its reliance on both individual and collective doses averted to population. The *economic criterion* is built up upon cost of spared doses by counter-measure implementation. Last criteria, more difficult to approach, are human ones, according to the social and psychological perception and acceptance of implemented countermeasures. In the work, we only take into account the radioecological point of view, based on reduction factors (RF) in agricultural products and the food chain.

Factors Influencing Counter-measures Effectiveness

Numbers of factors have an influence on the efficiency of counter-measures:

i) Radionuclides composition (half-life, type of fallout, mobility, physico-chemical form)

ii) Post-accidental time

iii) Environmental conditions (soils, landscape)

iv) Food practices and import/export ratio

v) Financial applicability

vi) Ecological acceptability

vii) Social perception

Efficiency of Counter-measures in Crop Production

Change of crops. For medium or long term, changes in land use, including crop selection, can be effective but with social or economical long term consequences. The most severe the change is the highest long-term potential consequences could be. That is why these measures should be implemented only for long life radionuclides when the others are not appropriated. The comparison cost/benefits has to be considered.

Removal of soil upper layer (0-10 cm). This measure is an effective technique for soil decontamination if its application does not deteriorate soil fertility and water regime. Efficiency can be very high (about 80%) according to the fact that most of the contamination, almost in the first time after accident, is in this upper layer. But the great disadvantages of the method are high procurement costs and problem of burial of a large body of radioactive wastes.

Ploughing. The efficiency depends on the type and depth of soil and on the type of crops, especially the rooting depth. Deep ploughing greatly reduces the uptake of radionuclide is recommended exclusively for soils with a powerful productive (humus horizon) layer (for example, chernozem and peaty soils).

Agrochemical measures. Effectiveness of liming is high but varying with initial soil pH. Average reduction factor for ^{90}Sr is 2 but could be higher (up to 10), for ^{137}Cs average 1.8 (up to 3). Liming reduces radiocaesium and radiostrontium uptake into plants by increasing the cation exchange capacity of the soil. It is usually inexpensive but easy to apply. Liming is the most effective in suppressing root uptake of radiocaesium when applied in conjunction with potassium. It is most effective when applied to organic soils. The application of potassium and organic fertilizer is inexpensive and comparatively easy to carry out. Due to the addition of potassium fertilizer ^{137}Cs intake in farm crops from different soils reduces on an average of 2 but up to 3-5 times. Effectiveness is high when applied to soil with low levels of available potassium.

The general table of the average reduction of accumulation of radionuclides for crop production counter-measures is given in Table 13.2.

Table 13.2. Average efficiency of crop production counter-measures.

Counter-measure	^{90}Sr	^{137}Cs
Use of mineral fertilizers	1.6	2.0
Liming	2.0	1.8
Ploughing (or other mechanical treatments)	1.4	1.4
Crop selection	4	4

For this category of counter-measures, the most efficient for ^{137}Cs is selection of crop species. The range is wide (0.54 to 20.3) but the mean value is about 4, which a very high level of efficiency. The selection of crop varieties is just after for its efficiency on ^{137}Cs accumulation. The less efficient methods are probably application of clay minerals and biological active matter.

For ^{90}Sr, the efficiency of the agricultural counter-measures is generally lower than for ^{137}Cs or at least at the same level, except for liming because of competitive action of calcium. The efficiency range is narrower than for ^{137}Cs. The most efficient method seams also to be the selection of crop varieties.

SUMMARY

In the soil environment, radionuclides can be dissolved in solution, held on inorganic soil constituents through various sorption or ion exchange reactions, complexed with soil organics, or precipitated as pure or mixed solids. Soluble contaminants are subject to migration with soil water, uptake by plants or aquatic organisms, or loss due to volatilisation.

Soil contaminated with radionuclides pose a long-term radiation hazard to human health through exposure via the food chain and other pathways. Remediation of radionuclide-contaminated soils has become increasingly important. Removal of the contaminated surface soil (often up to 40 cm) or immobilization of radionuclides in soils by applying mineral and chemical amendments are physically difficult and not likely cost-effective in practicality. Reducing plant uptake of radionuclides by competitive cations has the general advantage in large scale, low-level contamination incidents on arable land. There is very limited study on phytoextraction of radionuclides by specific plant species from contaminated sites.

14

Wastewater Irrigation and Soil Health

Wastewater irrigation is being practised in many countries around the world because of the substantial benefits accruing to communities as a result of reuse of nutrient-rich treated or untreated sewage. Treated effluent reuse has special significance for developing countries with arid and semi-arid climates and limited water resources. Land application of wastewater, industrial effluents and sludge provides nutrients for crop growth as well as organic matter for soil conditioning and it is often the most economic means of wastewater and sludge disposal. However, land application involves health and environmental risks, because sludge may contain heavy metals and industrial organic compounds, and wastewater and sludge may contain pathogens at detrimental concentrations.

SOIL AND GROUNDWATER POLLUTION

Discharge of untreated or partially treated wastewater to the soil gives rise to pollution problems in surface and groundwater and on land. Nitrates are a particular problem in many areas and the risk of contaminating groundwater through wastewater irrigation depend on local conditions as well as on the rate of application. Where a deep homogeneous unsaturated zone overlies the saturated layer of the aquifer, most pollutants is removed in the unsaturated layer and there is a very low risk of contaminating the groundwater. A high risk situation arise only in the case of a shallow and/or highly porous unsaturated zone above the aquifer, and especially if this zone is fissured.

Chemical pollutants. Municipal wastewater is likely to contain chemical pollutants wherever industrial discharges are allowed into the sewerage system. Of particular concern are those that are toxic to people, plants and aquatic biota. Heavy metals and refractory organics fall into this category. Boron, a constituent of synthetic detergents, is an important phytotoxin, and should be monitored when wastewater is used for irrigation. Preventing chemical pollutants from entering sewerage is the best solution but this is difficult to achieve unless industrial zones are isolated and provided with their own wastewater treatment plants.

A possible long-term problem with wastewater irrigation is build-up of toxic materials or salinity in the soil. As the unsaturated zone removes chemical pollutants, particularly heavy metals, their concentration in the soil increases with time and, after many years of irrigation, it is possible that toxic levels could develop and be taken up by a crop. The problem of soil salinization is common in arid regions, where irrigation water is saline and wastewater irrigation could give rise to this effect over the long-term, thereby rendering the land unusable for agriculture.

The productivity of irrigated land is fundamentally dependent on its internal drainage, which is a function of the soil profile morphology, pore size distribution and stability of pore structure. The first two factors are of paramount importance in relation to effluent reuse in irrigation. No irrigation scheme can succeed unless the soil profile remains permeable, and this depends both on the proportion of exchangeable cations held by the soil, other than exchangeable sodium percentage (ESP) and on the total concentration of soluble salts in the percolating water. Considerable evidence exists to indicate that pore structural stability is very important in determining the hydraulic properties of soils. The hydraulic conductivity of a soil is a function of the ESP and is related to the sodium adsorption ratio (SAR); the higher the SAR and the lower the electrolyte concentration of the percolating solution, the larger the hydraulic conductivity reduction. A further factor to consider in respect of wastewater irrigation is the high content of nutrients which might promote microbial growth, with consequent reduction in soil permeability and hydraulic conductivity.

Another aspect of agricultural concern is the effect of dissolved solids in the irrigation water on the growth of plants. Dissolved salts increase the osmotic potential of soil water and an increase in osmotic pressure of the soil solution increase the amount of energy which plants must expend to take up water from the soil. As a result, respiration is increased and the growth and yield of most plants decline progressively as osmotic pressure increases. Although, most plants respond to salinity as a function of the total osmotic potential of soil water, some plants are susceptible to specific ion toxicity. Many of the ions which are harmless or even beneficial at relatively low concentrations may become toxic to plants at high concentrations, either through direct interference with metabolic processes or through indirect effects on other nutrients, which might be rendered inaccessible.

The susceptibility of the population to long-term exposure to low levels of toxic chemicals, through the consumption of groundwater into which these materials have leached, is also of concern. Although, studies have indicated that only negligible amounts of such toxic chemicals normally move 30 cm beyond the point of application within the soil. It is possible that long term effluent reuse and eventual accumulation of toxic materials in the soil might lead ultimately to their mobilization and result in an increasing concentration showing up in groundwater.

Numerous studies have indicated that the content of certain toxic metals in plant tissues is directly proportional to the concentration of such metals within the soil root zone. Thus, long-term application of wastewater in irrigation poses the risk of plants having high levels of toxic materials in their tissues and the FAO recommends some maximum concentrations for phytotoxic elements in irrigation water.

Microbiological pollution. The potential risk of infection to humans, animals and plants from land application of treated wastewater is attributable to the presence of pathogenic organisms in the raw wastewater. It is clear from the many studies that, under favourable conditions, enteric pathogens can survive for extremely long periods of time on crops, in water or in the soil (Table 14.1). Factors that affect survival include the number and type of organisms, soil organic matter content, temperature, humidity, pH, amount of rainfall, amount of sunlight, protection provided by foliage, and competitive microbial flora. The survival of pathogens in soil has been reported to vary from a few hours to several months. Organisms such as Vibrio cholera have relatively short survival times, whereas other pathogens, including some bacterial species, ascaris ova and enteric viruses, appear to be highly resistant to environmental stress. It is thus evident that irrigation of health-sensitive crops (including fruits and vegetables) with raw or partially treated wastewater can present real health risks.

Table 14.1. Survival times of selected excreted pathogens in soil and on crop surfaces.

Pathogen	Survival time (days)	
	in soil	in crops
Viruses		
Enteroviruses	< 100 but usually < 20	< 60 but usually < 15
Bacteria		
Faecal coliforms	< 70 but usually < 20	< 30 but usually < 15
Salmonella spp.	< 70 but usually < 20	< 30 but usually < 15
Vibrio cholerae	< 20 but usually < 10	< 5 but usually < 2
Protozoa		
Entamoeba hystolytica cysts	< 20 but usually < 10	< 10 but usually < 2
Helminths		
Ascaris lumbricoides eggs	many months	< 60 but usually < 30
Hookworm larvae	< 90 but usually < 30	< 30 but usually < 10
Taenia saginata eggs	many months	< 60 but usually < 30
Trichuris trichiura eggs	many months	< 60 but usually < 30

Source: Feachem *et al.* (1983)

INTERACTION OF WASTEWATER WITH SOIL

The dissolved ions or compounds in wastewaters are retained in soil primarily through adsorption and chemical precipitation. Specific adsorption occurs when soils retain cations and anions so tightly that are not exchangeable. Heavy metals cation such as Zn, Cu, Ni, Cd and Pb, can be incorporated into soil via this mechanism. The application levels of Zn, Cu and Ni, which can be phytotoxic and, Cd, which can accumulate in plants to levels toxic to human/animals have been proposed according to relative toxicity of metals, soil texture, organic matter and cation exchange capacity.

The biological transformations of wastewater occur primarily in the root zone. The diversity of organisms in wastewaters enhances the soil's ability to degrade a wide variety of organic substances. The amount of oxygen present in soil largely determines the rate at which the organic matter decomposes and the end products that are formed.

WASTEWATER TREATMENT FOR AGRICULTURAL REUSE

Pre-application Treatment

Some degree of treatment must be provided to untreated municipal and industrial wastewater before it can be used for agricultural or landscape irrigation. The degree of pre-application treatment is a key factor in satisfactory operation and performance of a wastewater-soil-plant system. The required quality of effluent will depend on the crop or crops to be irrigated, the soil conditions and the system of irrigation adopted. Through crop restriction and selection of irrigation systems which minimize health risk, the level of wastewater treatment can be reduced. Adopting as low a level of treatment as possible is desirable, not only from the point of view of cost but also in acknowledgement of the difficulty of operating complex systems reliably. In many locations, it will be better to design the reuse system to accept a low-grade of effluent rather than rely on advanced treatment processes producing a reclaimed effluent which continuously meets a stringent quality standard. Nevertheless, there will be locations where a higher-grade effluent is necessary and it is essential that information on the performance of a wide range of wastewater treatment technology should be made available.

The design of wastewater treatment plants has usually been based on the need to reduce organic and suspended loads to limit pollution of the environment. Pathogen removal has very rarely been an objective. For reuse of effluents in agriculture, this must now be of primary concern and treatment processes should be selected and designed accordingly.

Removal of Microbiological Constituents

Conventional processes. Conventional wastewater treatment processes (plain sedimentation, activated sludge, biofiltration, aerated lagoons and oxidation ditches) are not effective, unless supplemented by disinfection, to produce an effluent which complies with the guideline for unrestricted irrigation.

Table 14.2. Expected removal of excreted bacteria, helminthes virus and cyst in various wastewater treatment processes

Treatment Process	Removal (10 log units) of			
	Bacteria	Helminths	Viruses	Cysts
Primary sedimentation				
Plain	0-1	0-2	0-1	0-1
Chemically assisted	1-2	1-3	0-1	0-1
Activated sludge	0-2	0-2	0-1	0-1
Biofiltration	0-2	0-2	0-1	0-1
Aerated lagoon	1-2	1-3	1-2	0-1
Oxidation ditch	1-2	0-2	1-2	0-1
Disinfection	2-6	0-1	0-4	0-3
Waste stabilization ponds	1-6	1- 3	1-4	1-4
Effluent storage reservoirs	1-6	1-3	1-4	1- 4

Source: Mara and Cairncross (1987)

Table 14.2 shows expected removal efficiencies in various wastewater treatment processes in respect of the major microbiological parameters of health concern.

Waste stabilization ponds. Ponds are usually the method of wastewater treatment of choice in warm climates wherever land is available at reasonable cost. Temperature is an important environmental factor affecting the efficiency of bacterial removal in stabilization ponds. A published WHO Manual on Stabilization Ponds (World Health Organisation, 1987) provides advice on planning, design and maintenance and emphasizes their low cost and simple operation. The only disadvantage of pond systems is the relatively large area of land that they require.

Soil-aquifer treatment. Soil-aquifer treatment offers the advantages of low-cost purification and requires much less land (approx. 0.5 to 1 m^2 per inhabitant) than stabilization ponds (5 m^2 per inhabitant for secondary ponds or 10 m^2 per inhabitant for ponds receiving raw sewage). This technique (incorporating in filtration basins) is adapted to permeable and aquifer soils.

Tertiary treatment. Disinfection (usually chlorination) of raw sewage has never been achieved in practice with full success. It can be used to reduce the numbers of excreted bacteria in the effluent from a conventional treatment plant if the plant is operating well. However, it is extremely difficult and costly to maintain a high, uniform and predictable level of disinfecting efficiency. In any case, chlorination will leave most helminth eggs totally unharmed. To reduce chlorination costs and assist in helminth removal, rapid-gravity sand filtration is often necessary as an additional tertiary treatment process. A more appropriate tertiary treatment option is to add one or more ponds in series to a conventional treatment plant. The addition of polishing ponds is a suitable measure to upgrade an existing wastewater treatment plant so that the effluent can be reused in irrigation of agricultural crops.

Removal of Inorganic Constituents

Conventional wastewater treatment. Conventional primary and secondary sewage treatment processes have a limited effect in removing inorganic components from sewage and will have only marginal effects on those effluent quality parameters of agricultural concern in reuse. The level of salinity will not be reduced significantly in passage through a conventional sewage treatment plant and the balance of sodium with other cations will not be changed to any extent. Physico-chemical treatment processes are more likely to be effective in the removal of inorganic constituents of wastewaters and may have to be considered where adverse environmental impacts will result from the long term use of saline effluent in irrigation. Primary settling removes a proportion of metals which are either insoluble or adsorbed onto particulate matter. Further metal removal occurs in the secondary biological stage of wastewater treatment, usually through adsorption of dissolved metals or fine particulate metals onto sludge flocs. It was found that for some metals (chromium, copper and lead) removal efficiency was greater in secondary treatment than in a primary process, while for zinc, the average removal percentage was similar at both stages. However, that metal removal at these stages shows great variability. The removal efficiency of advanced wastewater treatment processes can also be highly variable, with respect to both the process and the metal removes it.

Stabilization ponds. In recent years, there have been discussions on the utilization of phytoplanktonic algae, in algal ponds, for the removal of residual metals from wastewater. This technique is an economic method for removing heavy metals from wastewater, resulting in high quality effluent and valuable algal biomass which could be used for different purposes, one being the production of biogas. The combination of lime treatment and stabilization ponds might sometimes be an optimal treatment system when the ratio of sodium to calcium in sewage is not satisfactory.

Wastewater Application Control

Irrigation water, including treated wastewater, can be applied to the land in the five following general ways-

- by flooding (border irrigation), thus wetting almost all the land surface
- by furrows, thus wetting only part of the ground surface

- by sprinklers, in which the soil is wetted in much the same way as by rainfall,
- by subsurface irrigation, in which the surface is wetted little, if any, but the subsoil is saturated, and
- by localized (trickle or drip) irrigation, in which water is applied to each individual plant at an adjustable rate.

Sprinkler irrigation should not be used except for pasture or fodder crops and border irrigation should not be used for vegetables. Subsurface or localised irrigation can give the greatest degree of health protection, as well as using water more efficiently. However, it is expensive, and a high degree of reliable treatment is required to prevent clogging of the small holes (emitters) through which water is slowly released into the soil.

SUMMARY

Direct reuse of wastewater in the irrigation on the crops poses potential hazards and adverse impacts on soil environment. However, if appropriate low-cost technology for wastewater treatment and effluent distribution in irrigation can be developed to suit conditions and also provide the necessary safeguards to soil health, this form of reuse will conserve valuable water resources and increase crop production. Wastewater also undergoes natural physico-chemical and biological treatments in the soil matrix which provide not only a highly effective low-cost alternative to conventional treatment but also an ecologically balanced and environmentally compatible system of wastewater management.

References

Berti, W.R. and Jacob, L.W. (1996). Chemistry and phytotoxicity of soil trace elements from repeated sewage sludge application. J. Environ. Qual. *25:1025–1032.*

Blaylock, M.J., Salt, D.E., Dushenkhov, S., Zakharova, O., Gussman, C., Kapulnik, Y., Ensley, B.D. and Raskin, I. (1997). Enhanced accumulation of Pb in Indian mustard by soil-applied chelating agents. Environ. Sci. Technol. *31:860–865.*

Brooks, R.R. (1998). Plants that hyperaccumulate heavy metals. In Geobotany and hyperaccumulators. Wallingford, UK: CAB International, p. 55–94.

Chaudhry, T.M., Hayes, W.J., Khan, A.G. and Khoo, C.S. (1998). Phytoremediation-focusing on hyperaccumulator plants that remediate metal-contaminated soils. Aust. J. Ecotoxicol. *4:37–51.*

Chhonkar, P.K. and Rao, D.L. (1996). Soil biology and biochemistry in relation to land degradation and environmental pollution. Indian Society of Soil Science, *Bulletin No. 17. New Delhi.*

Cunningham, S.D., Berti, W.R. and Huang, J.W. (1995). Phytoremediation of contaminated soils. Tibitech, *13:393–397.*

Dhurvanarayana, V.V. (1993). Soil and water conservation Research in India. Indian Council of Agricultural Research, New Delhi.

Ebbs, D.S. and Kochian, L.V. (1997). Toxicity of zinc and copper to Brassica species: Implications for phytoremediation. J. Environ. Qual. *26:776–781.*

FAO (2001). Global estimates of gaseous emissions of NH_3, NO and N_2O from agricultural land, International Fertilizer Industry Association, Food and Agriculture Organization of the United Nations, Rome.

Feachem, R.G., Bradley, D.J., Garelick, H. and Mara, D.D. (1983). Sanitation and Disease: Health Aspects of Excreta and Wastewater Management, John Willey, Chichester.

Fitz, W.J. and Wenzel, W. W. (2002). Arsenic transformations in the soil rhizosphere-plant system: Fundamentals and potential application to phytoremediation. Journal of biotechnology, *99: 259-278.*

Fuentes, H.D., Khoo, C.S., Pe, T., Muir, S. and Khan, A.G. (2000). Phytoremediation of a contaminated mine site using plant growth regulators to increase heavy metal uptake. In: Sanches, M.A., Vergara, F., Castro, S.H., editors. Waste treatment and environmental impact in the mining industry. Victor Lamas, Concepcion, Chile, University of Concepcion Press. p. 427–35.

Hayes, W.J., Chaudhry, R.T., Buckney, R.T. and Khan, A.G. (2003). Phytoaccumulation of trace metals at the Sunny Corner mine, New South Wales, with suggestions for a possible remediation strategy. Aust. J. Ecotoxicol. *9:69–82.*

Hicks, Warren and Lamontagne, Sébastien (2006). A guide to sulphur gas emissions from wetlands and disposal basins: Implications for salinity management, CSIRO Land and Water Science Report 37/06, CRC LEME Open File Report 208.

Igwe, J.C., Nnorom, I.C. and Gbaruko, B.C. (2005). Kinetics of radionuclides and heavy metals behaviour in soils: Implications for plant growth. African Journal of Biotechnology *Vol. 4 (13), pp. 1541-1547.*

Iskandar, I.K. (1992). Restoration of metals contaminated soil. In: Environmental Restoration of metals contaminated soil, Lewis Publishers. p. 59.

Kabata-Pendias, A. (1992). Trace metals in soils in Poland-occurrence and behaviour, Soil Sci. *140:53-70.*

Khan, A.G. (2003). Vetiver grass as an ideal phytosymbiont for Glomalian fungi for ecological restoration of derelict land. In: Truong, P., Hanping, X., editors. Proceedings of the third international conference on vetiver and exhibition: Vetiver and water, Guangzou, China, October 2003. Beijing: China Agricultural Press; p. 466–74.

Khan, A.G. (2004). Mycotrophy and its significance in wetland ecology and wetland management. In: Wong, M.H., editor. Developments in ecosystems, vol. 1. Northhampton, UK: Elsevier, p. 97–114

Khan, A.G., Chaudhry, T.M., Hayes, W.J., Khoo, C.S., Hill, L., Fernandez, R. and Gallardo, P. (1998). Physical, chemical and biological characterization of a steelworks waste site at Port Kembla, NSW, Australia. Water Air Soil Pollut. *104: 389–402.*

Khan, A.G., Kuek, C., Chaudhry, T.M., Khoo, C.S. and Hayes W.J. (2000). Role of plants, mycorrhizae and phytochelators in heavy metal contaminated land remediation. Chemosphere., *41:197–207.*

Knox, A.S., Seamans, J.C., Mench, M.J., Vangronseveld, J. (2000). Remediation of metals and radionuclides. Contaminated soil using in situ stabilization Techniques. Macmillan Publishers, New York. pp.21-26.

Linger, P., Mussing, J., Fischer, H., Kobert, J. (2002). Industrial hemp (Cannabis sativa *L.) growing on heavy metal contaminated soil: Fibre quality and phytoremediation potential.* Ind. Crops Prod., *16:33–42.*

Lippmann, B., Leinhos, V. and Bergmann, H. (1995)). Influence of auxin producing rhizobacteria on root morphology and nutrient accumulation of crops. Angew Bot. *69:31–6.*

Mara, D.D. and Cairncross, A. (1987). Guidelines for the safe use of *wastewater and excreta in agriculture and aquaculture: Measures for public health protection, United Nations Environment Programme, World Health Organization, Geneva.*

Masle, J. and Passioura, J.B. (1987). The effect of soil strength on the growth of young wheat plants. Australian Journal of Plant Physiology, *14, 643–656.*

Mulligan, C.N., Young, R.N. and Gibbs, B.F. (1999). On the use of biosurfactants for the removal of heavy metals from oil contaminated soil. Environ. Prog.,*18(1):50–54.*

Mulligan, C.N., Young, R.N. and Gibbs, B.F. (2001). Remediation technologies for metal-contaminated soils and groundwater: An evaluation. Eng. Geol., *60:193–207.*

Passioura J.B. and Gardner P.A. (1990). Control of leaf expansion in wheat seedlings growing in drying soil. Australian Journal of Plant Physiology *17, 149–157.*

Passioura, J. B. (2002). Soil conditions and plant growth. Plant, Cell and Environment, *25, 311–318*

Passioura, J.B. and Stirzaker, R.J. (1993). Feedforward responses of plants to physically inhospitable soil. International Crop Science Congress, Ames, Iowa, 14–22 July 1992 (eds D.R. Buxton, R. Shibles, R.A. Forsberg, B.L. Blad, K.H. Asay, G.M. Paulsen & R.F. Wilson), Crop Science Society of America, *Madison, WI, USA.*

Patten, C.L. and Glick, B.R. (1996). Bacterial biosynthesis of indol-3 acetic acid. Can. J. Microbiol. *42:207–20.*

Patterson, J.W. (1985). Industrial wastewater treatment technology, 2nd ed. Boston: Butterworth

Pe, T., Fuentes, H.D., Khoo, C.S., Muir, S. and Khan, A.G. (2000). Preliminary experimental results in phytoremediation of a contaminated mine site using plant growth regulators to increase heavy metal uptake. In: Handbook and Abstracts, 15th Australian Statistical Conference, 3–7 July 2000. Adelaid Hilton International, South Australia, p. 143–4.

Romken, P., Bouwman, L., Japenga, J. and Draaisma, K. (2002). Potentials and drawbacks of chelate-enhanced phytoremediation of soils. Environ. Pollut.*116:109–21.*

Sethunathan, N. (1982). Biodegradation of agrochemicals in the soil environment, Review of Soil Research in India, Part I, 12th International Congress of Soil Science, 8-16 February 1982, New Delhi, India.

Shetty, K.G., Hetrick, B.A.D. and Schwab, A,P. (1995). Effects of mycorrhizae and fertilizer amendments on zinc tolerance of plants. Environ. Pollut. *88:307–14.*

Singh, A.K. (2007). Conversion to Organic Agriculture. International Book Distribution Co., Lucknow.

Singh, A.K. (2008). Production Technology on Bio-Organic Farm Inputs. International Book Distributing Co., Lucknow.

Singh, A.K. and Singh N.P. (2004). Agricultural Terminology, Concept Publisher, New Delhi.

Smedley, P.L. and Kinniburgh, D.G. (2002). A review of the source, behaviour and distribution of arsenic in natural waters, Applied geochemistry, *17: 517-568.*

Smith, L.A., Means, J.L., Chen, A., Alleman, B., Chapma, C.C., Tixier, Jr. J.S., Brauning, S.E., Gavaskar, A.R. and Royer, M.D. (1995). Remedial options for metal contaminated sites. Boca Raton, FL: Lewis

Stirzaker, R.J., Passioura, J.B. and Wilms, Y. (1996). Soil structure and plant growth: Impact of bulk density and biopores. Plant and Soil *185, 151–162.*

US Environmental Protection Agency (2000). Introduction to phytoremediation, Office of Research and Development. EPA/ 600/R-99/107.

Wenzel, W.W., Adriano, D.C., Salt, D. and Smith, R. (1999). Phytoremediation: A plant-microbe-based remediation system. In: Adriano, D.C. et al. editors. Bioremediation of Contaminated

Soils. Agronomy Monographs 37. Madison, WI: ASA, CSSA and SSSA, p. 457–508.

Yanez, L., Ortiz. D., Calderon, J., Batres, L., Carrizales, L. and Mejia, J. (2002). Overview of human health and chemical mixtures: problems facing developing countries. Environ. Health Perspect, *110 (Suppl. 6): 901–909.*

Appendices

Appendix I

LAND POLLUTION

Land pollution involves the following mechanism:

- Deposition of solid waste
- Accumulation of non-biodegradable materials
- Toxification of chemicals into poisons
- Alteration of soil chemical composition (imbalance of chemical equilibrium to soil medium)

Land pollution has a mass globally, everyday threatening the very foundation and mechanical support of every matter on earth. Statistically, it has been shown that:

- loss of 6 million hectares of land per year
- loss of 24 billion tons of topsoil per year
- loss of minimum 15 million acres prime agricultural land to overuse and mismanagement
- desertification of land results in the lost of 16 million per square miles of world's land surface

The causes for such devastation are generally due to 2 (two) forms of malpractices:

Unhealthy Soil Management Methods

- Improper tillage of soil in which excessive tillage result in the deterioration of soil structure.
- Non-maintenance of a proper supply of organic matter in the soil from the imbalance composition of the reserves of organic matter especially nitrogen, phosphorus and sulphur. Unplenished supply after cultivation of vegetation, living the soil prone to soil infertility, unable to stabilize the soil physicality which ultimately let to desertification.

- Irregular maintenance of a proper nutrient supply of trace elements gives rise to the use of excessive synthetic fertilizers, which are non-biodegradable and accumulate in the soil system which eventually destroys useful organisms such as bacteria, fungi and other organisms.
- Improper maintenance of the correct soil acidity which ultimately disrupts the adaptation of various crops and native vegetation of different soils as the solubility of minerals present, will be affected. In a more acidic soil, minerals tend to be more soluble and washed away during rainfall while alkaline soil, minerals are more insoluble which form complex minerals unable to be absorbed into the flora system physiological usage.

Improper Irrigation Practices

- Poorly drained soil result in salt deposits leading to high soil salinity that inhibit plant growth and may lead to crop failure.
- Unirrigated land giving rise to stagnation of agriculture waste products which accumulates and increases land toxicity.
- Irregular irrigation leads to decreasing moisturization of land for soil medium and replenishments of solvents for mineral.

Impacts of Land Pollution

Each kind of pollution has significant impacts to our everyday lives, affecting all living and non-living factors in the biosphere and the atmosphere and also involves socio-economic factors. These impacts have caused significant changes to the environment we are living in. Impacts can be seen from the following aspects:

- Biodiversity degradation
- Emergence of variant diseases
- Contamination of food leading to food shortage
- Economical effects

- Alteration to geographical landscapes
- Deforestation
- Desertification
- Erosion
- Landslide
- Climatic pattern change
- Global warming
- Distortion of food web
- Alteration to lifestyle due to overwhelming pollution

Appendix II

CLIMATIC PATTERN CHANGE: GLOBAL WARMING

Global Warming is the extraordinary increase of Earth's surface temperature due to the increase of greenhouse gases concentration on the atmosphere. Greenhouse gases are the heat-trapping gases in the atmosphere (carbon dioxide, methane, nitrous oxide, and CFC). They are the fundamental parts of the greenhouse effect, the role played by the atmosphere to continually warm the earth, by trapping some portions of heat that came from the solar energy (sun radiation) from reflecting back to space - just like the work of a greenhouse.

Causes of Global Warming

Rapid use of fossil fuel. Rapid use of fossil fuel will emit large amount of greenhouse gases, especially carbon dioxide.

Deforestation/clearing of lands. The increase of human population coincidentally demands for more lands to use. This leads to clearance of forest area in many regions. By eliminating forests, carbon dioxide that actually should be photosynthesized, is left in the atmosphere and accumulates.

The use of CFC in electronic appliances. CFC, or chlorofluorocarbon is a gas that decreases ozone (O_3) in the atmosphere, causing ozone depletion that increase the amount of solar radiation arriving to the Earth.

Open burning of trash. Open burning of trash worldwide emits greenhouse gases.

- Vehicle and industrial emissions
- Volcanic eruptions

Signs of Global Warming

- The receding of ice formations on Earth (snows at mountain-tops, glaciers, and Antarctic and Arctic ice)
- The increase of shrubbery in Arctic
- Thinner clouds over the sky, that decrease the ability to

reflect heat from the sun (as studied by NASA)

- The discovery of the decrease of Earth's albedo (the amount of sunlight reflection by the Earth surface to the Moon) by 2.5 percent, which means the Earth has loosen some levels of capability to reflect sunlight to the Moon.
- Change in wind directions

Impacts

- Stormy weather (more chances for hurricanes, floods, cyclones, and storms to happen).
- Increased severity for drought, hunger and spread of diseases, especially in poor countries.
- Declines of amphibians, caused by altered precipitation patterns resulted in lower levels of pond and lake waters, where amphibians survive.
- Damages to coral reefs
- Marine diseases
- Rising ocean temperature
- Ecosystem degradation
- Declining of biodiversity
- Economic and social downturn

Appendix III

LAND CLASSIFICATION

Land capability classification. The land-capability classification is an interpretative grouping of soils primarily meant for agricultural purposes and is mainly based on-

i. the inherent soil characteristics

ii. the external land features, and

iii. environmental factors.

The classification of soil units mapped into capability groupings enables one to understand-

i. the hazards of the soil to various factors which cause soil damage, deterioration in fertility, and

ii. its potentiality for production.

The capability classification provides 3 major categories-

i. a capability unit

ii. a capability sub-class

iii. capability class.

The capability unit. It is a grouping of soils that have almost the same influence on production and responses to the systems of management of common cultivated crops.

The capability sub-class. It is a grouping of capability units having similar limitations and hazards. Four kinds of limitations recognized are: (i) erosion, (ii) wetness, (iii) shallowness of root zone, and (iv) climate.

The capability class. It is the broadest category of classification. Eight (8) classes are recognized and the risk of soil damage or limitations in use progressively increases from class1 to class 8. Soils in classes 1 to 4 are arable. They are capable of producing commonly cultivated crops of the region under good management. Soils in the classes 5 to 7 are suited to adapted native plants, pasture or forestry. Soils of class 8 are neither suited for agriculture nor for silviculture.

Arable Land

Class I: These soils have only a few limitations that restrict their use. The soils are nearly level, deep, well drained with good water holding capacity. They are productive and suitable for intensive cropping.

Class II: these soils have some limitations that reduce the choice of crops and require moderate conservation practices to prevent deterioration when cultivated. The soils may be used for raising many of the cultivated crops.

Class III: These soils have severe limitations that reduce the choice of crops and require special conservation practices when used for raising cultivated crops. The limitations are more than in class II and thus need careful management.

Class IV: These soils have very severe limitations that reduce the choice of crops and need very careful management. The cultivation of crops may be restricted to once in 3-4 years. Conservation measures are more difficult to apply and maintain in these soils.

Non Arable Land

Class V: These soils have little or no erosion hazards but have other severe limitations such as wetness or overflow as in the case of bottom lands which prevent the normal tillage for common cultivated crops. These restrictions limit their use to pasture or adaptable tree species.

Class VI: These soils have severe limitations to depth, slope, erosion hazard etc. That makes them unsuitable for the cultivation of normal crops. Their use is restricted to pasture and silviculture.

Class VII: These soils have more limitations than those of class VI and make them totally unsuitable for cultivation of crops and restrict their use only to grazing.

Class VIII: These soils have severe limitations that preclude them from being used for agriculture or silviculture. Their use is restricted to recreation or wildlife.

Appendix IV

SOIL SAMPLING

Soil testing is a method for determining the fertility status of a soil, so that recommendations in regard to deficiency or toxicity of nutrients or soil amendments can be made. Soil testing laboratory is required to analyse quickly and recommendations made in respect of the fertilizer or soil amendment requirement for different soils or crops.

Soil tests and their interpretations are based on the samples analysed. Thus, the soils should be properly collected and should represent the area. Soil tests are as reliable as the samples drawn. For routine soil testing the field is traversed, variations in slope, color, texture, management and cropping patterns are noted, and it is divided into portions according to the variations. A separate samples are collected for each of the portions.

Samples are taken, using proper tools e.g. a soil tube, auger, spade or pickaxe for the plough depth at various spots and then composited. The soil is mixed thoroughly and spread on a sheet of paper or cloth and divided into 4 equal parts. Two opposite quarters are rejected and samples from the other two are mixed and this is repeated until the desired size of the sample is obtained (1/2 kg), collected in a plastic bag and placed in a box or cloth bag. Moist samples are dried before collecting. Care is taken to avoid contamination.

In the soil testing laboratory, soil samples are analysed by standardized rapid methods for the following items:

i. pH or the soil reaction which indicates whether the soil is acidic, alkaline or normal.

ii. Total soluble salts, as determined by electrical conductivity which indicates the degree of salinity, alkalinity of the soil.

iii. Organic carbon which is a measure of available nitrogen.

iv. Available phosphorus.

v. Available potassium.

The methods used are those standardized and tested for a particular area. In addition to the above items, other items like heavy metals (Zn, Cu, Mn, Cd, Ni, etc.) as considered necessary according to the problems, may be analysed. Based on soil analysis, the soils are classified into categories according to ratings given below:

Soil pH Interpretation

i.	Acidic	pH below 6.5
ii.	Normal to Saline	pH 6.6 - 8.5
iii.	Tending to become Alkaline	pH 8.6 - 9.0
iv.	Alkaline	pH above 9.0

Soil Fertility Interpretation

Parameters	Low	Medium	High
Organic carbon	below 0.5%	0.5 - 0.75%	above 0.75%
Available nitrogen (N)	below 280 kg/ha	280-560 kg/ha	above 560 kg/ha
Available phosphorus (P)	below 10 kg/ha	10-25 kg/ha	above 25 kg/ha
Available potassium (K)	below 110 kg/ha	110-280 kg/ha	above 280 kg/ha

Total soluble salts (conductivity in milli mhos/cm)

i.	below 1	normal
ii	1 - 2	critical for germination
iii.	2 - 4	critical for growth of sensitive crops
iv.	above 4	injurious to most crops

Procedure of Sampling

When? It should be done when there is not crop in the field i.e., before raising the next crop and also prior to application of manure or fertilizer. Rainy season should be avoided as far as possible for soil sampling.

Where? Ten to fifteen samples should be drawn from an apparently homogeneous plot. However, where the areas within a field look differently in appearance, crop growth, topography, drainage, soil type, cropping pattern, or are known to have been fertilized or manured differently, divide the field into parts/units and sample each portion (constituting a sampling unit) separately. There is no rigidity about the size of the sampling unit.

To what depth? As root penetration is important for crop growth, soil is sampled for cereal crops to a depth of 0-15 (plough layer). For deep rooted crops like sugarcane, cotton and horticultural crops, soil samples should be collected from deeper zones (15-30 cm and if necessary up to 60 cm) as well.

In case of saline-alkali or salt affected soils, salt crusts (visible or suspected) on the soil surface should be samples separately and the depth of sampling recorded.

What is composite sample? About 10 to 15 soil samples collected from a homogeneous sampling unit are pooled and intimately mixed and a smaller sample is drawn which should represent the entire sampling unit in its physical and chemical properties.

Materials required

i. Soil auger/soil tube/spade/pick-axe/Khurpi

ii. Bucket or tray

iii. Paper tags (labels)

iv. Information sheet

v. Cloth bags (alternatively polythene bags)

vi. Ball point pen or copying pencil.

Methods to be followed

Take a composite sample from each field as under:

- Remove the litter from the surface without much disturbing the soil.

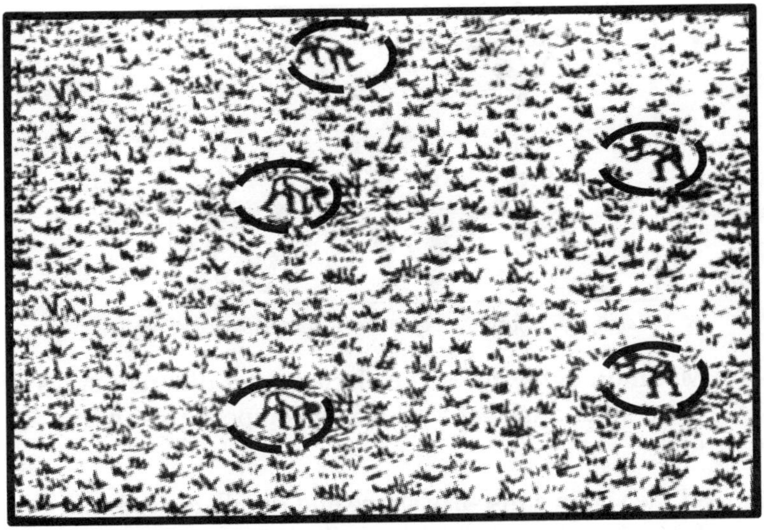

Removing of litter from few selected spots for sampling.

- Take soil samples from 10-15 random spots in a zigzag pattern in the field using a soil auger or by giving a 'V'-shaped cut up to depth of 15 cm with spade or khurpi at each spot and take 1.5 – 2.0 cm thick uniform slice of soil and collect them in a clean and dry container or bucket.

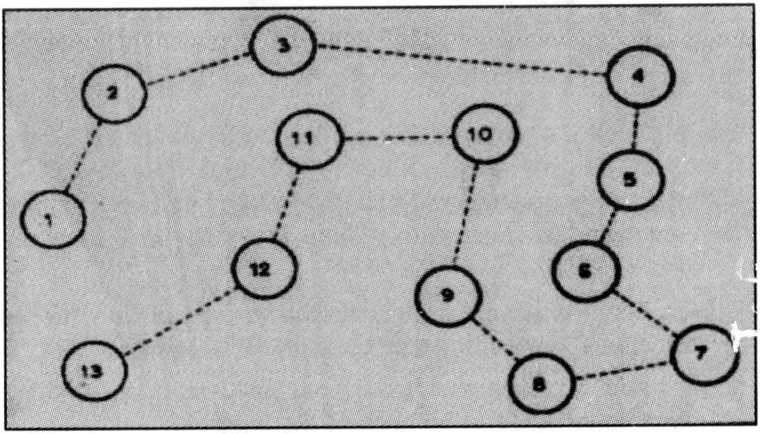

Zigzag Spots for sampling in a soil unit

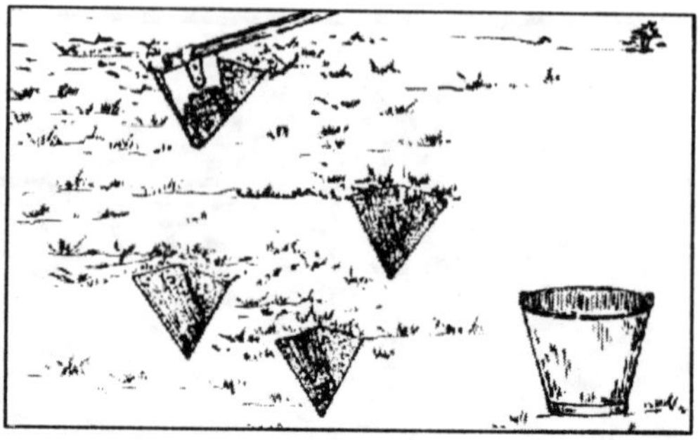

Taking of uniform slice of soil in V-shaped from each spot.

- Mix thoroughly the soil collected from different spots of the field and take at least 0.5 kg composite sample as follows:

Spread the soil in a disc-like shape and divide the soil collected into four parts (quartering) and discard one set of opposite quarters.

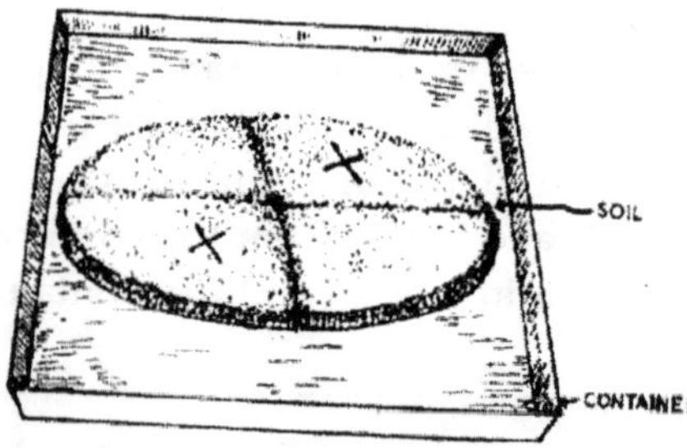

Dividing of collected soil mass into four parts and discard any two.

- Mix well the soil remaining two quarters and repeat this till the sample size is reduced to 0.5 to 0.75 kg (by taking only two parts) and put in a clean cloth bag free from any contamination of fertilizer, salts, etc. Polythene bags can also be used.
- Air-dry the sample under the shade. Do not dry the soil directly under sunlight or by artificial heating.
- Prepare two labels one to be put inside the bag and the other to be tied on the neck of the bag.
- One dully filled up soil information sheet should accompany with each sample.

Precautions to be taken

- Do not draw any sample from the extreme corners of the field, areas recently manure or fertilized, old bunds, marshy spots, manure piles and non-representative areas.
- Avoid sampling from furrows.
- Do not take less than 0.5 kg of composite sample.
- Keep the sample in a bag and tag it properly. See that each sample tag carries the name of the farmer, number of field or identify of the field, depth of the sample, date proceeding and succeeding crops and signature with name of the sampler. Use ballpoint pen or copying pencil for writing on the tag.
- Avoid use of fertilizer bags for handling of soil samples.
- Follow proper procedure for sampling in saline and alkaline soils as it is different from the procedure described here for normal soil conditions.

Appendix V

PERCENTAGE CHEMICAL COMPOSITION OF ROCKS

Oxides	Igneous rocks	Shales	Sandstones	Limestones
SiO_2	59.07	58.90	78.64	5.20
Al_2O_3	15.22	15.63	4.77	0.81
Fe_2O_3	3.10	4.07	1.08	0.54
MgO	3.45	2.47	1.17	7.92
FeO	3.71	2.48	0.32	--
CaO	5.10	3.15	5.51	42.74
Na_2O	3.71	1.32	0.45	0.05
K_2O	3.11	3.28	1.32	0.33
O_2	--	2.67	5.03	41.70
P_2O_5	0.30	0.17	.08	0.04
MnO	0.11	--	--	--
TiO_3	1.03	0.66	0.25	0.06
H_2O	1.30	3.72	1.33	0.56
Miscellaneous	0.79	1.48	0.07	0.05

Appendix VI

PARTICLE SIZES CLASSIFICATION SYSTEMS

Particle size	(Diameter in mm)	
	International system	USDA system
Gravel	2.0 and more	2.0 and more
Very coarse sand	2.0 –1.0	--
Coarse sand	1.0 - 0.5	2.0 - 0.2
Medium sand	0.5 - 0.25	--
Fine sand	0.25 – 1.0	0.2 - 0.02
Very fine sand	0.10 - 0.05	--
Silt	0.05 - 0.02	0.02 - 0.002
Clay	less than 0.002	less than 0.002

Appendix VII

CONVERSION FACTORS

Distance	1 sq. kilometre	=	100 hectares
	1 metre	=	1.0936 yards = 39.37 inches
	1 mile	=	1760 yards = 5,280 feet
	1 sq. mile	=	640 acres or 2.59 sq. km.
	1 yard	=	0.9144 metre (1 m = 1.094 yards)
	1 foot	=	0.3048 metre
	1 inch	=	2.54 centimetres (1 cm = 0.394 inch)
Area	1 acre	=	43,560 sq feet or 0.405 ha
	1 acre inch	=	3630 cubic feet or 102 cubic meters
	1 guntha	=	1/40th acre = 1089 sq ft = 33' * 33'
	1 acre	=	0.4047 hectare
	1 hectare	=	2.471 acres or 10,000 sq. metres
	1 hectare centimeter	=	100 cubic metre
	1 sq mile	=	640 acres
Weight	1 long ton	=	2,240 pounds
	1 short ton	=	2,000 pounds
	1 metric ton	=	1000 kilograms = 10 quintals
	1 long ton	=	28 maunds (approx.)
	1 metric ton	=	1.1023 short tons
	1 metric ton	=	0.9842 long ton
	1 short ton	=	0.9072 metric ton
	1 long ton	=	1.0161 metric ton
	1 pound	=	453.6 grammes = 0.4536 kg
	1 hundred-weight (cwt)	=	112 pounds = 50.8 kg
	1 kilogramme	=	2.2046 pounds
	1 lb per acre	=	1.12 kilogrammes per hectare
	1 cwt per acre	=	125.6 kilogrammes per hectare
Capacity	1 gallon (Imp.)	=	4.5461 litres
	1 gallon (U.S.)	=	3.7853 litres
	1 gallon (Imp.)	=	0.1604 cft
	1 litre	=	0.2200 gallon (Br.)
	1 litre	=	2642 gallon(U.S.)
	1 bushel	=	56 lb (potatoes)
	1 bushel	=	56 lb (barley)
	1 bushel	=	60 lb wheat
	1 bushel	=	42 lb (oats)
	1 gallon water weighs	=	10 lb
	1 cft of water weighs	=	62.3 lb

Appendix VIII

CONVERSION FACTORS DETERMINING QUANTITIES OF FERTILIZERS

Quantity	Multiplied by-	Gives corresponding quantity of-
Nitrogen	4.854	Ammonium sulphate
Nitrogen	2.222	urea
Nitrogen	3.846	Ammonium sulphate nitrate
Nitrogen	4.000	Ammonium chloride
Nitrogen	3.030	Ammonium nitrate
Phosphoric acid (P_2O_5)	6.250	Superphosphate, single
Phosphoric acid (P_2O_5)	12.222	Superphosphate, double
Phosphoric acid (P_2O_5)	2.857	Dicalcium phosphate
Phosphoric acid (P_2O_5)	5.000	Bone meal, raw
Potash (K_2O)	1.666	Muriate of potash
Potash (K_2O)	2.000	Sulphate of potash
Ammonium sulphate	.206	Nitrogen
Sodium nitrate	0.155	Nitrogen
Urea	0.450	Nitrogen
Ammonium sulphate nitrate	0.260	Nitrogen
Ammonium chloride	0.250	Nitrogen
Ammonium nitrate	0.330	Nitrogen
Superphosphate, double	0.450	Phosphoric acid (P_2O_5)
Dicalcium phosphate	0.350	Phosphoric acid (P_2O_5)
Bone meal, raw	0.200	Phosphoric acid (P_2O_5)
Muriate of potash	0.600	Potash (K_2O)
sulphate of potash	0.500	Potash (K_2O)

Appendix IX

PERIODIC TABLE OF ELEMENTS

H Hydrogen 1.00794																	**He** Helium 4.003
3 **Li** Lithium 6.941	4 **Be** Beryllium 9.012182											5 **B** Boron 10.811	6 **C** Carbon 12.0107	7 **N** Nitrogen 14.00674	8 **O** Oxygen 15.9994	9 **F** Fluorine 18.9984032	10 **Ne** Neon 20.1797
11 **Na** Sodium 22.989770	12 **Mg** Magnesium 24.3050											13 **Al** Aluminum 26.981538	14 **Si** Silicon 28.0855	15 **P** Phosphorus 30.973761	16 **S** Sulfur 32.066	17 **Cl** Chlorine 35.4527	18 **Ar** Argon 39.948
19 **K** Potassium 39.0983	20 **Ca** Calcium 40.078	21 **Sc** Scandium 44.955910	22 **Ti** Titanium 47.867	23 **V** Vanadium 50.9415	24 **Cr** Chromium 51.9961	25 **Mn** Manganese 54.938049	26 **Fe** Iron 55.845	27 **Co** Cobalt 58.933200	28 **Ni** Nickel 58.6934	29 **Cu** Copper 63.546	30 **Zn** Zinc 65.39	31 **Ga** Gallium 69.723	32 **Ge** Germanium 72.61	33 **As** Arsenic 74.92160	34 **Se** Selenium 78.96	35 **Br** Bromine 79.904	36 **Kr** Krypton 83.80
37 **Rb** Rubidium 85.4678	38 **Sr** Strontium 87.62	39 **Y** Yttrium 88.90585	40 **Zr** Zirconium 91.224	41 **Nb** Niobium 92.90638	42 **Mo** Molybdenum 95.94	43 **Tc** Technetium (98)	44 **Ru** Ruthenium 101.07	45 **Rh** Rhodium 102.90550	46 **Pd** Palladium 106.42	47 **Ag** Silver 107.8682	48 **Cd** Cadmium 112.411	49 **In** Indium 114.818	50 **Sn** Tin 118.710	51 **Sb** Antimony 121.760	52 **Te** Tellurium 127.60	53 **I** Iodine 126.90447	54 **Xe** Xenon 131.29
55 **Cs** Cesium 132.90545	56 **Ba** Barium 137.327	57 **La** Lanthanum 138.9055	72 **Hf** Hafnium 178.49	73 **Ta** Tantalum 180.9479	74 **W** Tungsten 183.84	75 **Re** Rhenium 186.207	76 **Os** Osmium 190.23	77 **Ir** Iridium 192.217	78 **Pt** Platinum 195.078	79 **Au** Gold 196.96655	80 **Hg** Mercury 200.59	81 **Tl** Thallium 204.3833	82 **Pb** Lead 207.2	83 **Bi** Bismuth 208.98038	84 **Po** Polonium (209)	85 **At** Astatine (210)	86 **Rn** Radon (222)
87 **Fr** Francium (223)	88 **Ra** Radium (226)	89 **Ac** Actinium (227)	104 **Rf** Rutherfordium (261)	105 **Db** Dubnium (262)	106 **Sg** Seaborgium (263)	107 **Bh** Bohrium (262)	108 **Hs** Hassium (265)	109 **Mt** Meitnerium (266)	110 (269)	111 (272)	112 (277)	113	114				

58 **Ce** Cerium 140.116	59 **Pr** Praseodymium 140.90765	60 **Nd** Neodymium 144.24	61 **Pm** Promethium (145)	62 **Sm** Samarium 150.36	63 **Eu** Europium 151.964	64 **Gd** Gadolinium 157.25	65 **Tb** Terbium 158.92534	66 **Dy** Dysprosium 162.50	67 **Ho** Holmium 164.93032	68 **Er** Erbium 167.26	69 **Tm** Thulium 168.93421	70 **Yb** Ytterbium 173.04	71 **Lu** Lutetium 174.967
90 **Th** Thorium 232.0381	91 **Pa** Protactinium 231.03588	92 **U** Uranium 238.0289	93 **Np** Neptunium (237)	94 **Pu** Plutonium (244)	95 **Am** Americium (243)	96 **Cm** Curium (247)	97 **Bk** Berkelium (247)	98 **Cf** Californium (251)	99 **Es** Einsteinium (252)	100 **Fm** Fermium (257)	101 **Md** Mendelevium (258)	102 **No** Nobelium (259)	103 **Lr** Lawrencium (262)

Glossary

A

Abiotic. Not biotic or living.

Absorption. Uptake of matter or energy by a substance.

Acid rain. Acid rain forms when certain atmospheric gases (primarily carbon dioxide, sulphur dioxide, and nitrogen oxides) come in contact with water in the atmosphere or on the ground and are chemically converted to acidic substances. Oxidants play a major role in several of these acid-forming processes. Carbon dioxide dissolved in rain is converted to a weak acid (carbonic acid). Other gases, primarily oxides of sulphur and nitrogen, are converted to strong acids (sulphuric and nitric acids). Although, rain is naturally slightly acidic because of carbon dioxide, natural emissions of sulphur and nitrogen oxides, and certain organic acids, human activities can make it much more acidic.

Acid soil. Soil with a pH value less than 7.0.

Acidification. Process whereby soil becomes acid ($pH<7$) because acid parent material is present or in regions with high rainfall, where soil leaching occurs. Acidification can be accelerated by human activities (use of fertilizers, deposition of industrial and vehicular pollutants).

Acidophile. Organism that grows best under acid conditions (down to a pH of 1).

Active ingredient. In any pesticide product, the component that kills, or otherwise controls, target pests. Pesticides are regulated primarily on the basis of active ingredients.

Activity. The amount of a radioactive substance expressed as the number of disintegrations taking place per unit time, or as Curies or Bequerel. Count rate is a non-absolute measure of activity dependent on the counting efficiency.

Adaptation. Change in an organism or population of organisms through which they become more suited to the prevailing environment. Adaptation can be genetic and/or physiological.

Adsorption. Process by which atoms, molecules or ions are retained on the surfaces of solids by chemical or physical bonding.

Aeration of soil. Amount of air-filled pores in the soil, expressed as the volume difference between total porosity and actual soil moisture. Optimum soil aeration is 30% but strongly depends on the structure and packing state of soil particles; 15-20% is normally satisfactory for the growth of grasses and cereals; below 10% is not good for plant growth.

Aerobe. An organism that can grow in the presence of air.

Aerobic. An environment that has a partial pressure of oxygen similar to normal atmospheric conditions.

Afforestation. The act or process of establishing a forest, especially on land not previously forested.

Aggregation. Process whereby primary soil particles (sand, silt, clay) are bound together, usually by natural forces and substances derived from root exudates and microbial activity. Soil aggregates are arranged to form soil peds, units of soil structure, classified by size, shape (platy, prismatic, columnar, angular, subangular, blocky, granular, etc.) and grade (single-grain, massive, weak, moderate, strong). From an agronomical point of view, the most important soil aggregates are in range 3 - 1 mm.

Agricultural waste. Grain, vegetable, and fruit harvest residue, poultry and livestock manure, and residual materials in liquid or solid form generated from the production and marketing of poultry, livestock, and their products.

Air pollution. The existence in the air of substances in concentrations that are determined unacceptable to human health and the environment. Contaminants in the air we breathe, come mainly from manufacturing industries, electric power plants, exhaust from automobiles, buses and trucks.

Air quality standards. The level of selected pollutants set by law that may not be exceeded in outside air. Used to determine the amount of pollutants that may be emitted by industry.

Algae. Any of various primitive, chiefly aquatic, one- or multi-celled, non-flowering plants that lack true stems, roots, and leaves, but usually contain chlorophyll. Algae convert carbon dioxide and inorganic nutrients, such as nitrogen and phosphorus, into organic matter through photosynthesis and form the basis of the marine food chain. Common algae include dinoflagellates, diatoms, seaweeds, and kelp.

Algal bloom. A condition which occurs when excessive nutrient levels and other physical and chemical conditions facilitate rapid growth of algae. Algal blooms may cause changes in water colour. The decay of the algal bloom may reduce dissolved oxygen levels in the water.

Alkalophile. Organism that grows best under alkaline conditions (up to a pH of 10.5).

Anaerobic. Any process that can occur without molecular oxygen; also applicable to organisms that can survive without free oxygen.

Anion exchange capacity. Sum of exchangeable anions that a soil can adsorb. Usually expressed as centimoles, or millimoles, of charge per kilogram of soil (or of other adsorbing material such as clay).

Anion. *Particle with a negative charge. See also* ion, cation.

Anoxic. *Literally* 'without oxygen'. *An adjective describing a habitat devoid of oxygen.*

Aquifer. A subsurface water-bearing layer that will yield water in a usable quantity to a well or spring.

Aquitard. A subsurface semi-confining layer that can store ground water and also transmit it slowly from one aquifer to another.

Arable land. Agricultural land that is cultivated by ploughing, usually to 20 or 30 cm depth. More than 30 cm represents deep ploughing.

Asbestos. A mineral fibre that can pollute air or water and cause cancer or asbestosis when inhaled. Use of asbestos has been banned or severely restricted in manufacturing and construction.

Assimilative capacity. The ability of a natural body of water to receive wastewaters or toxic materials without harmful effects and without damage to aquatic life.

Atmosphere. The air surrounding the Earth, described as a series of layers of different characteristics. The atmosphere, composed mainly of nitrogen and oxygen with traces of carbon dioxide, water vapour and other gases, acts as a buffer between Earth and the sun. The layers, troposphere, stratosphere, mesosphere, thermosphere and the exosphere, vary around the globe and in response to seasonal changes.

Autotroph. *An organism that uses carbon dioxide as its source of carbon for growth.* (*Compare with* Heterotroph).

B

Bacteria. A group of diverse and ubiquitous prokaryotic single-celled microorganisms.

Bactericide. A pesticide used to control or destroy bacteria, typically in the home, schools, or on hospital equipment.

Bedload. Sediment or other material that slides, rolls, or bounces along the streambed or channel bed of flowing water.

Benthic. Living in or on the bottom of a body of water.

Benthos. Collectively, all organisms living in, on, or near the bottom substrate in aquatic habitats (examples are oysters, clams, burrowing worms).

Best management practices (BMPs). Management practices (such as nutrient management) or structural practices (such as terraces) designed to reduce the quantities of pollutants - such as sediment, nitrogen, phosphorus, and animal wastes - that are washed by rain and snow melt from farms into nearby receiving waters, such as lakes, creeks, streams, rivers, estuaries, and groundwater.

Bioaccumulation. Intracellular accumulation of environmental pollutants such as heavy metals by living organisms.

Bioassay. A method of testing a material's effects on living organisms.

Bioaugmentation. The addition to the environment of microorganisms that can metabolize and grow on specific organic compounds.

Bioavailability. The availability of chemicals to potentially biodegradative microorganisms.

Biochemical oxygen demand (BOD). The requirement for molecular oxygen by microbes during oxidation of biological substances in sewage. The BOD test measures the oxygen consumed (in mg/L) over 5 days at 20 degrees C.

Biochemicals. Chemicals that are either naturally occurring or identical to naturally occurring substances. Examples include hormones, pheromones, and enzymes. Biochemicals function as pesticides through non-toxic, non-lethal modes of action, such as disrupting the mating pattern of insects, regulating growth, or acting as repellants. Biochemicals tend to be environmentally compatible and are thus important to Integrated Pest management programs.

Biodegradable. The ability of a substance to be broken down physically and/or chemically by microorganisms. For example, many chemicals, food scraps, cotton, wool, and paper are biodegradable; plastics and polyester generally are not.

Biodiversity. The number and variety of different organisms in the ecological complexes in which they naturally occur. Organisms are organized at many levels, ranging from complete ecosystems to the biochemical structures that are the molecular basis of heredity. Thus, the term encompasses different ecosystems, species, and genes that must be present for a healthy environment. A large number of species must characterize the food chain, representing multiple predator-prey relationships.

Biofilter. Apparatus that biodegrades volatile organic compounds in air by passing the air through media containing biodegrading microbes.

Biological pesticides. Certain microorganisms, including bacteria, fungi, viruses, and protozoa that are effective in controlling target pests. These agents usually do not have toxic effects on animals and people and do not leave toxic or persistent chemical residues in the environment.

Biomass. The amount of living matter present in a particular habitat.

Biome. Well-defined terrestrial environment (e.g. desert, tundra, or tropical forest). The complex of living organisms found in an ecological region.

Bioremediation. The use of living organisms (e.g., bacteria) to clean up oil spills or remove other pollutants from soil, water, and wastewater.

Biosphere. Part of the Earth system in which life can exist, between the outer portion of the geosphere and the inner portion of the atmosphere.

Biostimulation. A process that increases activity of microorganisms biodegrading contaminants. For example, addition of nutrients, oxygen, or other electron donors and acceptors.

Biota. All living organisms in a given area.

Biotransformation. Alteration of thc structure of a compound by a living organism or enzyme.

Bioventing. The process of supplying oxygen in situ to oxygen deprived soil microbes by forcing air through unsaturated contaminated soil at low flow rates. This stimulates biodegradation and minimizes stripping volatiles into the atmosphere. Frequently used to remediate soil under structures since it is relatively non-invasive.

Bog. Wetland that has no significant inflows or outflows, supports acidophilic mosses, particularly Sphagnum and in which peat is accumulating. Similar to fen, marsh, pocosin, swamp, and wetland.

Boulder clay. Unstratified glacial deposits laid down directly beneath the ice or dropped from the surface as the ice melted; boulder clay and till are synonymous terms for this unsorted material which ranges from rock flour to rocks and boulders of great size, according to the nature of the bedrock.

Brownfield. An abandoned, idled, or under-used industrial or commercial facility where expansion or redevelopment is complicated by a real or perceived environmental contamination.

BTEX. Benzene, toluene, ethylbenzene, and xylenes.

C

Calcification. Process whereby the soil is kept sufficiently supplied with calcium to saturate the soil cation exchange sites.

Capillary water. Water in capillary pores influenced by forces that hold water in soils against a tension usually greater than 60cm. Capillary water can move upwards against gravity.

Carbon cycle. Sequence of transformations whereby carbon dioxide is converted to organic forms by photosynthesis or chemosynthesis, recycled through the biosphere (with partial incorporation into sediments), and ultimately returned to its original state through respiration or combustion.

Carcinogenic. Capable of causing cancer. A suspected carcinogen is a substance that may cause cancer in humans or animals but for which the evidence is not conclusive.

Cation exchange. Interchange between a cation in solution and another cation in the boundary layer between the solution and surface of negatively charged material such as clay or organic matter.

Cation. Particle with positive charge; reactions between anions and cations create electrical forces.

Chemical oxygen demand (COD). The amount of oxygen in

milligrams per liter to oxidize both organic and oxidizable inorganic compounds.

Chlorination. Adding chlorine to water or wastewater, generally for the purpose of disinfection, but frequently for accomplishing other biological or chemical results. Chlorine also is used almost universally in manufacturing processes, particularly for the plastics industry.

Chlorofluorocarbons (CFCs). A family of chemicals commonly used in air conditioners and refrigerators as coolants and also as solvents and aerosol propellants. CFCs drift into the upper atmosphere where their chlorine components destroy ozone. CFCs are thought to be a major cause of the ozone hole over Antarctica.

Chronic effect. An adverse effect on any living organism in which symptoms develop slowly over a long period of time or recur frequently.

Clay coating/film. Coatings of oriented clay on the surfaces of peds and mineral grains and lining pores, also called clay skins, clay flows, illuviation cutans, or argillans.

Clay minerals. Clay-sized hydrous aluminium silicates having a large interlayer space that can hold significant amounts of water and other substances; they have large surface area allowing swelling and shrinking; examples are montmorillonite or smectite and kaolinite.

Clay. Soil particle smaller than 0.002mm or 2μm, with high specific area mainly influencing soil colloidal properties (see also colloid) as well as stability of soil structure. high stability in both wet and dry conditions; also a soil texture class.

Climate Change. This term is commonly used interchangeably with *'global warming'* and *'the greenhouse effect'*, but is a more descriptive term. Climate change refers to the build-up of man-made gases in the atmosphere that trap the suns heat, causing changes in weather patterns on a global scale. The effects include changes in rainfall patterns, sea level rise, potential droughts, habitat loss, and heat stress. The greenhouse gases of most concern are carbon dioxide, methane, and nitrous oxides.

Clouds. A visible mass of liquid droplets suspended in the atmosphere above Earth's surface. Clouds form in areas where air rises and cools. The condensing water vapour forms small droplets of water (0.012 mm) that, when combined with billions of other droplets, form clouds. Clouds can form along warm and cold fronts, where air flows up the side of the mountain and cools as it rises higher into the atmosphere, and when warm air blows over a colder surface, such as a cool body of water.

Coating. Layer of a substance completely or partly covering a surface of soil material; coatings can comprise clay, calcite, gypsum, iron, organic material, salt, etc.

Coliform bacteria. See Fecal coliform bacteria.

Colloid. Particle, which may be a molecular aggregate, with a diameter of 0.1 to 0.001µm; clay and soil organic matter are often called soil colloids because they have particle sizes that are within, or approach, colloidal dimensions.

Colluvial. Pertaining to material or processes associated with transportation and/or deposition by mass movement (direct gravitational action) and local, unconcentrated runoff on slopes and/or at the base of slopes.

Colluvium. Unconsolidated, unsorted colluvial material.

Cometabolism. The biodegradation of a pollutant by an organism while using some other compound(s) for growth and energy. There is little or no benefit to the biodegrading organism, the pollutant just happens to be affected by the growth of the cometabolizing organism.

Commercial waste. All solid waste from businesses. This category includes, but is not limited to, solid waste originating in stores, markets, office buildings, restaurants, shopping centers, and theaters.

Compost. Decomposed organic material that is produced when bacteria in soil break down garbage and biodegradable trash, making organic fertilizer. Making compost requires turning and mixing and exposing the materials to air.

Concentration. The relative amount of a substance mixed with another substance. An example is five parts per million of carbon monoxide in air or 1 milligram/litre of iron in water.

Condensation. Change of a substance to a denser form, such as gas to a liquid. The opposite of evaporation.

Conservation tillage. Any tillage and planting system that maintains at least 30% of the soil surface covered by residue after planting for the purpose of reducing soil erosion by water.

Conservation. Preserving and renewing natural resources to assure their highest economic or social benefit over the longest period of time. Clean rivers and lakes, wilderness areas, a diverse wildlife population, healthy soil, and clean air are natural resources worth conserving for future generations.

Construction and Demolition waste. Waste building materials, dredging materials, tree stumps, and rubble resulting from construction, remodelling, repair, and demolition operations on houses, commercial buildings and other structures, and pavements. May contain lead, asbestos, or other hazardous materials.

Contaminant. See Pollutant.

Contour. An imaginary line on the surface of the earth connecting points of the same elevation. A line drawn on a map connecting points of the same elevation.

Controlled drainage. The use of surface and subsurface drainage and control structures to control the water table depth in a field.

Convection. The rising of warm air and the sinking of cool air. When a layer of air receives enough heat from the Earth's surface, it expands and moves upward. Colder, heavier air flows under it which is then warmed, expands and rises. The warm rising air cools as it reaches higher, cooler regions of the atmosphere and begins to sink. Convection causes local breezes, winds, and thunderstorms.

Corrosive. A substance that eats or wears away materials gradually by chemical action.

Creosote. An antifungal wood preservative used frequently to treat telephone poles and railroad ties. Creosote consists of coal tar distillation products, including Phenols and PAHs.

D

DCE. Dichloroethylene.

Decalcification. Removal of calcium carbonate or calcium ions from the soil by leaching.

Deep well injection. A process by which waste fluids are injected deep below the surface of the earth.

Denitrification. The process by which nitrate-nitrogen is converted to nitrogen gas by soil microorganisms when soil oxygen is low or absent.

Deposition. The settling out of a soil particle or aggregate of particles from the water column.

Designated use. A beneficial use type established by a state for each water resource and specified in water quality standards, whether or not it is being attained.

Detachment. The process of a soil particle, nutrient, or pesticide breaking free from its position in the soil.

Detritus. Fragments of plant material.

Diagnostic horizon. Dominant soil property or material defines name of horizon, e.g. gypsic horizon having distinct calcium sulphate (gypsum. $CaSO_4$) enrichment; genetic horizon depending on the type of pedogenesis.

Diagnostic horizon. *see* horizon.

Diatoms. Any number of microscopic algae whose cell walls consist of two box-like parts or valves and contain silica.

Dinoflagellates. Unicellular biflagellate algae with thick cellulose plates.

Discharge. The release of any waste into the environment from a point source. Usually refers to the release of a liquid waste into a body of water through an outlet such as a pipe, but also refers to air emissions.

Disposal facility. A landfill, incinerator, or other facility which receives waste for disposal. The facility may have one or many disposal methods available for use. Does not include wastewater treatment.

Disposal. The discharge, deposit, injection, dumping, spilling, leaking, or placing of any solid waste or hazardous waste into the environment (land, surface water, ground water, and air).

Dissolved oxygen. The amount of oxygen present in the water column. More than 5 parts oxygen per million parts water is considered healthy; below 3 parts oxygen per million is generally stressful to aquatic organisms.

Dose response. How an organism's response to a toxic substance changes as its overall exposure to the substance changes. For example, a small dose of carbon monoxide may cause drowsiness; a large dose can be fatal.

Dose. In terms of monitoring exposure levels, the amount of a toxic substance taken into the body over a given period of time.

Drainage area. An area of land that drains to one point; watershed.

Drainage ditch. A ditch that is constructed in a field in order to allow agricultural fields to drain more quickly.

Dump. A land site where wastes are discarded in a disorderly or haphazard fashion without regard to protecting the environment. Uncontrolled dumping is an indiscriminate and illegal form of waste disposal. Problems associated with dumps include multiplication of disease-carrying organisms and pests, fires, air and water pollution, unsightliness, loss of habitat, and personal injury.

E

Earth system. The Earth regarded as a unified system of interacting components, including geosphere (land), atmosphere (air), hydrosphere (water and ice), and biosphere (life).

Ecology. The study of the relationships between all living organisms and the environment, especially the totality or pattern of interactions; a view that includes all plant and animal species and their unique contributions to a particular habitat.

Ecosystem. The interacting synergism of all living organisms in a particular environment; every plant, insect, aquatic animal, bird, or land species that forms a complex web of interdependency. An action taken at any level in the food chain, use of a pesticide for example, has a potential domino effect on every other occupant of that system.

Electrical conductivity (EC). Conduction of electricity through water or a solution of soil commonly used to estimate the soluble salt content in solution, e.g. soil solution.

Electron acceptor. Small inorganic or organic compound that is reduced to complete an electron transport chain. Compound that is reduced in a metabolic redox reaction.

Electron donor. Small inorganic or organic compound that is oxidized to initiate an electron transport chain. Compound from which electrons are derived in a metabolic redox reaction.

Emission standards. Government standards that establish limits on discharges of pollutants into the environment (usually in reference to air).

Emission. The release or discharge of a substance into the environment. Generally refers to the release of gases or particulates into the air.

Endangered species. Animals, plants, birds, fish, or other living organisms threatened with extinction by man-made or natural changes in the environment.

Energy recovery. To capture energy from waste through any of a variety of processes (e.g., burning). Many new technology incinerators are waste-to-energy recovery units.

Enhanced rhizosphere biodegradation. Enhanced biodegradation of contaminants near plant roots where compounds exuded by the roots increase microbial biodegradation activity. Other plant processes such as water uptake by the plant roots can enhance biodegradation by drawing contaminants to the root zone.

Enrichment. Culture in a liquid medium that results in the increase of the population of an organism relative to others. The liquid culture frequently contains substances that encourage the growth of the selected organism (such as the chemical pollutant and mineral nutrients.)

Environmental equity. Equal protection from environmental hazards for individuals, groups, or communities regardless of race, ethnicity, or economic status.

Enzymes. Proteins that act as biological catalysts.

Erosion. The wearing away of the land surface by water, wind, ice, gravity or other natural or anthropogenic agents that abrade, detach and remove soil particles or rock material from one point on the earth's surface, for deposition elsewhere, including gravitational creep and so-called tillage erosion.

Estuary. A complex ecosystem between a river and near-shore ocean waters where fresh and salt water mix. These brackish areas include bays, mouths of rivers, salt marshes, wetlands, and lagoons and are influenced by tides and currents. Estuaries provide valuable habitat for marine animals, birds, and other wildlife.

Eutrophication. A process by which a water body becomes rich in dissolved nutrients, often leading to algal blooms, low dissolved oxygen, and changes in community composition. Eutrophication occurs naturally, but can be accelerated by human activities that increase nutrient inputs to the water body.

Ex Situ. Out of the original position (excavated).

Exposure. Radiation or pollutants that come into contact with the body and present a potential health threat. The most common routes of exposure are through the skin, mouth, or by inhalation.

Extremely Hazardous Substances (EHS). Any of 366 (+ or -) chemicals or hazardous substances identified by EPA on the basis of hazard or toxicity and listed under EPCRA. The list is periodically revised.

F

Facultative organism. Organism that can carry out both options of a mutually exclusive process (e.g., aerobic and anaerobic metabolism).

Fecal coliform. Bacteria from the colons of warm-blooded animals which are released in fecal material.

Feldspar. Group of hard crystalline minerals that consist of aluminum silicates of sodium or calcium or barium.

Fen. Flat and swampy land, usually low in altitude and similar to a bog or marsh.

Fermentation. An energy yielding metabolism that involves a series of oxidation-reduction reactions in which the substrate and terminal electron acceptor are organic compounds. Fermentation occurs in a wide variety of bacteria and fungi.

Fertilization. Application of mainly mineral compounds, in order to increase soil fertility. In some cases, (e.g. liming) the purpose of fertilization is also to improve specific soil properties (pH, stability of soil structure).

Field capacity. Field capacity has been defined as the soil moisture state when, 48 hours after saturation or heavy rain, all downward movement of water has ceased. It is the water content retained at low suctions (5-33kPa) depending on soil type, and is the upper limit of plant available water.

Fine texture. A broad group of textures consisting of, or containing, large quantities of fine fractions, particularly silt and clay. Includes sandy clay, silty clay, and clay texture classes. When used in reference to family particle-size classes, is specifically defined as having 35 to 60 percent clay. *See also soil texture.*

Fluvioglacial deposits. Material moved from the margins of glaciers and subsequently sorted and deposited by streams flowing from the melting ice.

Forested riparian buffer. Streamside forest that is used to control non-point source pollution and in particular, sediment and nutrients.

Fugitive emissions. Air pollutants released to the air other than those from stacks or vents; typically small releases from leaks in plant equipment such as valves, pump seals, flanges, sampling connections, etc.

Fungicide. A pesticide used to control or destroy fungi on food or grain crops.

G

Garbage. Food waste (animal and vegetable) resulting from the handling, storage, packaging, sale, preparation, cooking, and serving of foods.

Geomorphology. Science of landforms that studies the evolution of the Earth's surface and interprets landforms as records of geological history.

Geosphere. Physical elements of the Earth's surface, crust and interior. Processes in the geosphere include continental drift, volcanic eruptions and earthquakes.

Gibbsite. $Al(OH)_3$. Mineral with a platy structure, that occurs in highly weathered soils and in laterite.

Glacial drift. Unstratified deposits laid down directly beneath the ice or dropped from the surface as the ice melted.

Glaciers. Large masses of ice that form by the compaction and recrystallization of snow under freezing conditions; glaciers often move downslope or outward in all directions because of the stress of their own weight; they may be stagnant or retreating under warming conditions.

Glaciofluvial deposits. Material moved by glaciers and subsequently sorted and deposited by streams flowing from the melting ice. The deposits are stratified and may occur in the form of outwash plains, deltas, kames, eskers, and kame terraces. *See also Glacial drift and till.*

Glaciolacustrine deposits. Material ranging from fine clay to sand derived from glaciers and deposited in glacial lakes by water originating mainly from the melting of glacial ice; many such deposits are bedded or laminated with varves.

Gley soil. Soil formed under naturally wet or waterlogged conditions as evidenced by grey colours stemming from the reduction, under anaerobic conditions, of ferric iron to the ferrous state.

Global warming. See definition for Climate Change.

Greenhouse effect. Process by which significant changes in the chemistry of Earth's atmosphere may enhance the natural process that warms our planet and elevates temperatures. Certain gaseous components of the atmosphere, called greenhouse gases, transmit the visible portion of solar radiation but absorb specific spectral bands of thermal radiation emitted by the Earth. The theory is that terrain absorbs radiation, heats up, and emits longer wavelength thermal radiation that is prevented from escaping into space by the blanket of carbon dioxide and other greenhouse gases in the atmosphere. As a result, the climate warms. See *Climate change.*

Greenhouse gas. A gaseous component of the atmosphere contributing to the greenhouse effect. Greenhouse gases are transparent to certain wavelengths of the sun's energy, allowing them to penetrate deep into the atmosphere or all the way into the Earth's surface. Greenhouse gases and clouds prevent some of infrared radiation from escaping, trapping the heat near the

Earth's surface where it warms the lower atmosphere. Alteration of this natural barrier of atmospheric gases can raise or lower the mean global temperature of the Earth. Greenhouse gases include carbon dioxide, methane, nitrous oxide, chlorofluorocarbons (CFC's) and water vapour. Carbon dioxide, methane and nitrous oxide have significant natural and human sources while only industries produce CFC's. Water vapour has the largest greenhouse effect, but its concentration in the troposphere is determined within the climate system. Water vapour will increase in response to global warming, which in turn may further enhance global warming.

Groundwater- Water found below the surface of the land, usually in porous rock formations. Groundwater is the source of water found in wells and springs and is used frequently for drinking.

Gully. Channel resulting from erosion and caused by the concentrated but intermittent flow of water during and immediately following heavy rainfall; gullies are deep enough (usually >0.5 m) to interfere with, but not obliterated by, normal tillage operations.

H

Halflife. The time it takes for any initial amount of a radionuclide to be halved by radioactive decay.

Halophilic. Organisms whose requirement for salt exceeds that of other organisms.

Heavy metal. A common hazardous waste; can damage organisms at low concentrations and tends to accumulate in the food chain.

Herbicide. A pesticide designed to control or kill plants, weeds, or grasses. Almost 70% of all pesticide used by farmers and ranchers are herbicides. These chemicals have wide-ranging effects on non-target species (other than those the pesticide is meant to control).

Heterotroph. Any organism that requires exogenous organic material for growth and reproduction.

Horizon. Single layer in soil profile with similar properties or material but which differs at least in one property, e.g. colour or texture from adjacent horizons above or below in the profile.

Household or Domestic waste. Solid waste composed of garbage and rubbish, which normally originates from residential, private households, or apartment buildings. Domestic waste may contain a significant amount of toxic or hazardous waste from improperly discarded pesticides, paints, batteries, and cleaners.

Humification. Process whereby the carbon of organic residues is transformed and converted to humic substances through biochemical and abiotic processes.

Humus. Organic compounds in soil, exclusive of undecayed plant and animal tissues, their partial decomposition products, and the soil biomass; a term often used synonymously with soil organic matter, its structure is amorphous, specific weight is low and surface area high. Humus is important for soil fertility, and helps to bind soil particles and aggregates together.

Hydrocarbons. Chemicals that consist entirely of hydrogen and carbon. Hydrocarbons contribute to air pollution problems like smog.

Hydromorphic soils. Formed under conditions of poor drainage in marshes, swamps, seepage areas or flats.

Hydrosphere. Water and ice on or near the surface of the Earth. Includes water vapour in clouds; ice caps and glaciers; and water in the oceans, rivers, lakes and aquifers. Processes in the hydrosphere include the flow of rivers, evaporation and rain.

I

Impaired waters. Surface and ground waters that are negatively impacted by pollution resulting in decreased water quality.

In Situ. In place, without excavation.

Incineration. The destruction of solid, liquid, or gaseous wastes by controlled burning at high temperatures. Hazardous organic compounds are converted to ash, carbon dioxide, and water. Burning destroys organics, reduces the volume of waste, and vaporizes water and other liquids the wastes may contain. The residue ash produced may contain some hazardous material, such as non-combustible heavy metals, concentrated from the original waste.

Industrial waste. Unwanted materials produced in or eliminated from an industrial operation and categorized under a variety of headings, such as liquid wastes, sludge, solid wastes, and hazardous wastes.

Inoculum. Material used to introduce a microorganism into a suitable situation for growth.

Insecticide. A pesticide compound specifically used to kill or prevent the growth of insects.

Ion. Electrically charged atom or group of atoms.

Irradiated food. Food that has been briefly exposed to radioactivity (usually gamma rays) to kill insects, bacteria, and mold. Irradiated food can be stored without refrigeration or chemical preservatives and has a long *'shelf life'*.

Isoenzyme. An enzyme that occurs in more than one form in a given species. Sometimes called an isozyme.

Isotopes. Nuclides belonging to the same chemical but differing in mass, that is same atomic number but different neutron number.

K

Karst. Topography with sinkholes, caves and underground drainage that is formed in limestone, gypsum or other rocks by dissolution (dissolving).

L

Lake. A man-made impoundment or natural body of freshwater of considerable size, whose open-water and deep-bottom zones (no light penetration to bottom) are large compared to the shallow-water (shoreline) zone, which has light penetration to its bottom.

Land treatment. The whole range of BMPs implemented to control or reduces non-point source pollution.

Land use. The way land is developed and used in terms of the types of activities allowed (agriculture, residences, industries, etc.) and the size of buildings and structures permitted. Certain types of pollution problems are often associated with particular land uses, such as sedimentation from construction activities.

Landfill. A method for final disposal of solid waste on land. The refuse is spread and compacted and a cover of soil applied, so that effects on the environment (including public health and safety) are minimized. Under current regulations, landfills are required to have liners and leachate treatment systems to prevent contamination of ground water and surface waters. An industrial landfill disposes of non-hazardous industrial wastes. A municipal landfill disposes of domestic waste including garbage, paper, etc. This waste may include toxins that are used in the home, such as insect sprays and powders, engine oil, paints, solvents, and weed killers.

Landslide. A general term for a mass movement landform and a process characterized by moderately rapid to rapid (greater than 30 cm per year) downslope transport by means of gravitational stresses, of a mass of rock and regolith that may or may not be water saturated.

Leaching. Removal of soluble materials from one zone in soil to another via water movement in the profile.

Lethal dose. A dose of ionising radiation sufficient to cause death. Median lethal dose (MPLD or ld-50) is dose required to kill within a specified period of time (usually 30 days) half of the individuals in a large group of organisms similarly exposed.

The LD-50/30 for man is about 400-500 roentgens.

Lignin. A complex polymer that occurs in woody material of higher plants. It is highly resistant to chemical and enzymatic degradation. The white rot fungi are known for their lignin degrading capability.

Litter. The highly visible portion of solid waste (usually packaging material) which is generated by the consumer and carelessly discarded outside of the regular garbage disposal system, as on the highways or in streets.

Loading. The influx of pollutants to a particular water body.

Loess. Material transported and deposited by wind and consisting of predominantly silt-sized particles, forming important fertile soils.

M

Marsh. A transition zone between water and land usually covered by grass.

Mass movement. Dislodgement and downslope transport of soil and rock material as a unit under direct gravitational stress; includes slow displacements, such as creep and rapid movements such as landslides, rock slides, earthflows, debris flows and avalanches; water, ice and to a lesser extent, air usually play an important role in the process.

Maximum contaminant level (MCL). The maximum level of certain contaminants permitted in drinking water supplied by a public water system as set under Safe Drinking Water Act.

Mesophile. An organism whose optimum growth range is 20-45 degrees C. (Compare with Thermophile and Psychrophile.)

Methanogen. Bacteria that anaerobicly oxidize hydrogen to methane and water using carbon dioxide as the electron acceptor. These occur in anaerobic muds, ponds, and sewage sludge.

Methanotroph. Aerobic bacteria that can use methane as a sole source of carbon.

Microaerophilic. An environment that is low in oxygen but is not anaerobic.

Microcosm. A community or other unit that is representative of a larger unity.

Microflora. All of the microorganisms associated with location or environment.

Micronutrient. Chemical element necessary for growth found in small amounts, usually < 100 mg/kg in a plant. These elements consist of B, Cl, Cu, Fe, Mn, Mo, and Zn.

Microorganisms. Bacteria, yeasts, simple fungi, algae, protozoans, and a number of other organisms that is microscopic in size. Most are beneficial but some produce disease. Others are involved in composting and sewage treatment.

Milligrams/liter (mg/l). A measure of concentration used in the measurement of fluids. mg/l is the most common way to present a concentration in water and is roughly equivalent to parts per million.

Mineralization. The conversion of humus and soil organic matter into inorganic substances by microbial breakdown.

Moderately fine textured. Texture group consisting of clay loam, sandy clay loam and silty clay loam textures. See also *soil texture.*

Monolith. Representative vertical section taken from vertical face of a soil profile pit or section, which represents arrangement of soil horizons; there are various methods of how to take and conserve soil monoliths.

Munsell colour system. Colour designation system that specifies the relative degrees of the three simple variables of colour i.e. hue (wavelength), value (degree of lightness or darkness), and chroma (purity or strength). For example, 10YR 6/4 is a colour (of soil) with a hue = 10YR, value = 6, and chroma = 4.

Mutagenicity. The property of a chemical that causes the genetic characteristics of an organism to change in such a way that future generations are permanently affected.

Mycelium. (plural, mycelia). Mass of hyphae that form the vegetative body of many fungal organisms.

Mycobacterium. A genus of aerobic bacteria found in soil and water that are capable of biodegrading multi-ring compounds such as PAHs.

Mycorrhiza. A mutually beneficial association between a fungus and the root of a plant. These occur in a wide range of plants including trees, shrubs, and herbaceous plants.

N

Natural isotope. Naturally occurring nuclide of any given chemical element. A natural isotope is either stable, or very long life, or the descendent of a very long-lived forefather.

Neurotoxin. A poisonous compound that acts on the nervous system.

Nitrate Respiration. (dissimilatory nitrate reduction) The use of nitrate as a terminal electron acceptor for anaerobic respiration. This process occurs under anaerobic or microaerophilic conditions. Not all bacteria are capable of this form of metabolism and the nitrate may not be reduced completely to nitrogen gas (stopping at nitrite, for example).

Nitrate. A form of nitrogen which is readily available to both aquatic and terrestrial plants as a nutrient.

Nitrification. The oxidation of ammonia to nitrite and then nitrate by bacterial species such as *Nitrosomonas* and *Nitrobacter*, respectively. This process is strictly aerobic.

Nitrogen. An element that is a component of protein structures in living organisms.

Non-point source controls. General phrase used to refer to all methods employed to control or reduce non-point source pollution.

Non-point source pollution. Pollution originating from runoff from diffuse areas (land surface or atmosphere) having no well-defined source.

No-till. The practice of leaving the soil undisturbed from harvest to planting except for nutrient injection. Planting or drilling is accomplished in a narrow seedbed or slot created by coulters, row cleaners, disk openers, in-row chisels, or rototillers. Weed control is accomplished primarily with herbicides.

Nutrient management. A BMP designed to minimize the contamination of surface and groundwater by limiting the amount of nutrients (usually nitrogen) applied to the soil to no more than the crop is expected to use. This may involve changing fertilizer application techniques, placement, rate, or timing. The term fertilizer includes both commercial fertilizers and manure.

Nutrients. Chemicals that are needed by plants and animals for growth (e.g., nitrogen, phosphorus). In water resources, if other physical and chemical conditions are optimal, excessive amounts of nutrients can lead to degradation of water quality by promoting excessive growth, accumulation, and subsequent decay of plants, especially algae. Some nutrients can be toxic to animals at high concentrations.

O

Obligate. Any state or condition that is an essential attribute of a given organism. For example, an obligate aerobe can grow only under aerobic conditions.

Oligotrophic. Bodies of water poor in those nutrients that support growth of aerobic photosynthetic organisms.

Organic pump. Uptake of large quantities of water by plant (trees) roots and translocation into the atmosphere to reduce a flow of water. Used to keep contaminated groundwater from reaching a body of water, or to keep surface water from seeping into a capped landfill and forming leachate.

Organic soil material. Consists of organic debris that accumulates at the surface under either wet or dry conditions and in which any mineral component present does not significantly affect the soil properties. Organic soil material must have organic carbon (organic matter) contents as follows. (i) if saturated with water for long periods (unless artificially drained), and excluding live roots, either. 18% organic carbon (30% organic matter) or more if the mineral fraction comprise 60% or more clay; or 12% organic carbon (20% organic matter) or more if the mineral fraction has no clay; or a proportional lower limit of organic carbon content between 12 and 18% if the clay content of the mineral fraction is between 0 and 60%; or (ii) if never saturated with water for more than a few days, 20% or more organic carbon.

Organic soil. A soil in which the sum of the thickness of layers comprising organic soil materials is generally greater than the sum of the thickness of mineral layers.

Organically grown. Food, feed crops, and livestock grown within an intentionally-diversified, self-sustaining agro-ecosystem. In practice, farmers build up nutrients in the soil using compost, agricultural wastes, and cover crops instead of synthetically derived fertilizers to increase productivity, rotate crops, weed mechanically, and reduce dramatically their dependence on the entire family of pesticides. Farmers must be certified to characterize crops as organically grown and can only use approved natural and synthetic biochemical, agents, and materials for three consecutive years prior to harvest. Livestock must be fed a diet that includes grains and forages that have been organically grown and cannot receive hormones, sub-therapeutic antibiotics, or other growth promoters.

Organism. Any living being, whether plant, mammal, bird, insect, reptile, fish, crustacean, aquatic or estuarine animal, or bacterium.

Oxidant. A substance containing oxygen that reacts chemically with other materials to produce new substances. Oxidants are the primary ingredients in smog.

Oxidase. An enzyme that catalyses a reaction in which electrons are removed from a substrate and donated directly to molecular oxygen.

Oxygenase. An enzyme that catalyses a reaction in which one (monooxygenase) or both (dioxygenase) atoms of molecular oxygen are incorporated into a molecule of substrate.

Oxygenases catalyze the first step in degradation of strait-chained and aromatic hydrocarbons.

Ozone hole. A large area of intense stratospheric ozone depletion over the Antarctic continent that typically occurs annually between late August and early October, and generally ends in mid-November. This severe ozone thinning has increased conspicuously since the late seventies and early eighties. This phenomenon is the result of chemical mechanisms initiated by man-made chlorofluorocarbons (CFC's). Continued build-up of CFC's is expected to lead to additional ozone loss worldwide. The frigid air permits the formation of ice clouds that facilitate chemical interactions among nitrogen, hydrogen and chlorine (elevated from CFC's) atoms, the end product of which is the destruction of ozone.

Ozone layer. The layer of ozone that begins approximately 15 km above Earth and thins to an almost negligible amount at about 50 km, shields the Earth from harmful ultraviolet radiation from the sun. The highest natural concentration of ozone (approximately 10 ppm by volume) occurs in the stratosphere at approximately 25 km above Earth. The stratospheric ozone concentration changes throughout the year as stratospheric circulation changes with the seasons. Natural events such as volcanoes and solar flares can produce changes in ozone concentration, but man-made changes are of the greatest concern.

Ozone. An almost colourless, gaseous form of oxygen with an odour similar to weak chlorine. A relatively unstable compound of three atoms of oxygen, ozone constitutes on the average less than 1 ppm ozone concentration in the atmosphere (peak ozone concentration in the stratosphere can get as high as 10 ppm). Yet ozone in the stratosphere absorbs nearly all of the biologically damaging solar ultraviolet radiation before it reaches the Earth's surface where it can cause skin cancer, cataracts, and immune deficiencies and can harm crops and aquatic ecosystems. Ozone is produced naturally in the middle and upper stratosphere through dissociation of molecular oxygen by sunlight. In the absence of chemical species produced by human activity, a number of competing chemical reactions among naturally-occurring species, primarily atomic oxygen, molecular oxygen, and oxides of hydrogen and nitrogen maintains the proper ozone balance. In the present-day stratosphere, this natural balance has been altered, particularly by the introduction of man-made chlorofluorocarbons. If the ozone decreases, the ultraviolet radiation at the Earth's surface will increase. Tropospheric ozone is a by-product of the photochemical (light-induced) processes associated with air pollution. See photochemical smog. Ozone in the troposphere can damage plants and humans.

P

PAH. Polynuclear aromatic hydrocarbon. Multi-ring compounds found in fuels, oils, and creosote. These are also common combustion products.

Parent material. Mineral or rock material on and/or from which soils are formed during pedogenesis (soil formation process); parent material is one of the five major soil forming factors.

Parts per million (ppm). A volume unit of measurement; the number of parts of a substance in a million parts of another substance. For example, 10 ppm nitrate in water means 10 parts of nitrate in a million parts of water.

Pasture. Grassland used for grazing of mainly domestic herbivores.

Pathogen. An organism capable of causing disease.

Peat. Organic soil material with more than 50% of organic matter derived from plant residues with not fully destroyed structure. Peat forms in a wet soil environment or below the water table where mineralisation of organic matter comes close to zero; a peat horizon or layer is normally more than 30cm thick.

Peatland. A generic term for any wetland where partially decayed plant matter accumulates; mire, moor and muskeg are terms used for peatlands in Europe and Canada; see also *bog* and *fen*.

Pedogenesis. Process of soil formation and development by soil forming factors. climate (mainly temperature and precipitation), parent material, living organisms (plants and biota), topography, time, water and Man.

Pedon. A three-dimensional body of soil with lateral dimensions (1 to 10 m^2) large enough to permit the study of horizon shapes and relations.

Perennial stream. A watercourse that flows throughout the year or most of the year (90%), in a well defined channel. Same as a *live stream.*

Periglacial. Pertaining to processes, conditions, areas, climates and topographic features occurring at the immediate margins of glaciers and ice sheets and influenced by cold temperature of the ice.

Permafrost. (i) permanently frozen subsurface material underlying the solum; (ii) perennially frozen soil horizon where temperature remains below 0°C throughout the year and in which Cryosols form.

Permanent grassland. Natural (mainly steppe areas) or agricultural soils with grass cover not normally ploughed.

Pesticide management. A BMP designed to minimize contamination of soil, water, air, and non-target organisms by controlling the amount, type, placement, method, and timing of pesticide application necessary for crop production.

Pesticide. Substances intended to repel, kill, or control any species designated a *'pest'* including weeds, insects, rodents, fungi, bacteria, or other organisms. The family of pesticides includes herbicides, insecticides, rodenticides, fungicides, and bactericides.

pH. The measure of acidity or alkalinity of a chemical solution, from 0 to 14. Anything neutral, for example, has a pH of 7. Acids have a pH less than 7, bases (alkaline) greater than 7.

Phenol. Carbolic acid (C_6H_5OH). Phenols and substituted phenols are used as antimicrobial agents in high concentrations.

Phosphorus. An element essential to the growth and development of plants, but which, in excess, can cause unhealthy conditions that threaten aquatic animals in surface waters.

Phytoaccumulation. *See* Phytoextraction.

Phytodegradation. A process in which plants are able to degrade (break down) organic pollutants through their metabolic processes.

Phytoextraction. Use of plants to extract contaminants (such as metals) from the environment (especially soil). When the plants are saturated with contaminants they are harvested.

Phytomining. Use of plants to extract inorganic substances of economic value (precious metals, etc.).

Phytoremediation. Use of plants to remediate contaminated soil or groundwater.

Phytostabilization. Use of soil amendments and plants to reduce bioavailability and offsite migration of contaminants.

Phytovolatilization. Use of plants to volatilise contaminants (solvents, etc.) from soil or water.

Plasmid. Extra DNA in a cell that is usually dispensable, but may confer an advantage to the cell, such as the ability to biodegrade certain compounds or resistance to antibiotics.

Ploughing (tillage). Mechanical cultivation of agricultural soils by the plough to different depths (20-30cm), creating arable land.

Plume. A concentration of contaminants in air, soil, or water usually extending from a distinct source.

Pocosin. A bog formed in shallow depressions with poor drainage, supporting predominantly evergreen shrubs or small trees.

Point source. Any confined and discrete conveyance from which pollutants are or may be discharged. These include pipes, ditches, channels, tunnels, conduits, wells, containers, and concentrated animal feeding operations.

Pollutant. A contaminant that adversely alters the physical, chemical, or biological properties of the environment. The term includes nutrients, sediment, pathogens, toxic metals, carcinogens, oxygen-demanding materials, and all other harmful substances. Pollutant means dredged spoil, solid waste, incinerator residue, sewage, garbage, sewage sludge, munitions, chemical wastes, biological materials, radioactive materials, heat, wrecked or discarded equipment, rock, sand, cellar dirt, and industrial, municipal, and agricultural waste discharged into water.

Pollution prevention. Actively identifying equipment, processes, and activities which generate excessive wastes or use toxic chemicals and then making substitutions, alterations, or product improvements. Conserving energy and minimizing wastes are pollution prevention concepts used in manufacturing, sustainable agriculture, recycling, and clean air/clean water technologies.

Primary mineral. A mineral that has not been altered chemically since crystallization and deposition from molten lava. See also *secondary mineral.*

Protection of soil. Conscious process necessary for soil and soil properties preservation realised at different levels (personal, local, national, continental) and using information obtained by soil research. Sustainability is the result of this process.

Psychrophile. An organism with an optimum growth temperature less than 20 degrees C.

R

Radioactivity. The spontaneous decay or disintegration of an unstable atomic nucleus accompanied by the emission of ionising radiation.

Radioactive decay. The spontaneous transformation of one nuclide into a different nuclide or into a different energy state of the same nuclide. The process results in decrease, with time, of the number of original radioactive atoms in a sample. It involves the emission of alpha particles, beta particles, gamma rays or the nuclear capture or ejection of orbital electrons, of fission. Also called radioactive disintegration.

Radioactive waste. Any waste that emits energy as rays, waves, or streams of energetic particles. Radioactive materials are often mixed with hazardous waste, usually from nuclear reactors, research institutions, or hospitals.

Radon. A colourless, naturally occurring gas formed by radioactive decay of radium atoms. Radon accumulating in basements and other areas of buildings without proper ventilation has been identified as a leading cause of lung cancer.

Rain forest. An evergreen woodland of the tropics distinguished by a continuous leaf canopy and an average rainfall of about 300 cms per year. Rain forests play an important role in the global environment. Destruction of tropical rain forests reduces the amount of leaf area in the tropics, and consequently the amount of carbon dioxide absorbed, causing increases in levels of carbon dioxide and other atmospheric gases. It is estimated that cutting and burning of tropical forests contributes about 20 percent of the carbon dioxide added to the atmosphere each year. Rain forest destruction also means the loss of a wide spectrum of biological life, erosion of soil, and possible desertification.

Recalcitrant. Resistant to biodegradation.

Recycling. Reusing materials and objects in original or changed forms rather than discarding them as wastes.

Redox Potential. The oxidation-reduction potential of an environment. Measures the tendency of the environment to be reducing (donate electrons) or oxidizing (accept electrons).

Reduction. Gain of an electron by an atom, ion, or molecule; a decrease in the oxidation state.

Reductive Dechlorination. Removal of Cl from an organic compound by reducing the carbon atom from C-Cl to C-H.

Reference Dose (RfD). The particular concentration of a chemical that is known to cause health problems. A standard that also may be referred to as the acceptable daily intake.

Regolith. The unconsolidated mantle of weathered rock and soil material on the Earth's surface, sometimes considered to be loose earth materials above solid rock.

Release. Any spilling, leaking, pumping, pouring, emitting, emptying, discharging, injecting, escaping, leaching, dumping, or disposing into the environment of a hazardous or toxic chemical, or extremely hazardous substance.

Resource recovery. The extraction of useful materials or energy from solid waste. Such materials can include paper, glass, and metals that can be reprocessed for re-use. Resource recovery also is employed in pollution prevention.

Respiration. Energy yielding metabolism in which oxygen is the terminal electron acceptor for substrate oxidation.

Rhizofiltration. Uptake of contaminants by the roots of plants immersed in water. When the roots are saturated with contaminants they are harvested.

Rhizoplane. The surface of plant roots.

Rhizosphere. Soil in the area surrounding plant roots that is influenced by the plant root. Typically a few millimeters or at most centimeters from the plant root. Important because this area is higher in nutrients and thus has a higher and more active microbial population.

Riffles. Areas of a stream or river characterized by a rocky substrate and turbulent, fast-moving, shallow water.

Riparian. Relating to the bank or shoreline of a body of water.

River. A watercourse that flows at all times, receiving ground or surface water, for example, from other streams or rivers. The terms *'river'* and *'stream'* are often used interchangeably, depending on the size of the water resources and the region in which they are located.

Rodenticide. A pesticide or other agent used to kill rats and other rodents or to prevent them from damaging food, crops, or forage.

Root exudates. Substances released from plant root system in drops or small quantities of carbohydrates, organic acids, vitamins and many other substances essential for life of soil microorganisms.

Runoff. Water that is not absorbed by soil but rather drains off the land into bodies of water, either via surface or subsurface flows.

S

Saline soil. A non-sodic soil (see sodic soil) containing sufficient soluble salt to adversely affect the growth of most crop plants. The lower limit of electrical conductivity in the saturation extract of such soils is conventionally set at 4 dS m^{-1} (at 25°C), though sensitive plants are affected at about half of this salinity and highly tolerant ones at about twice of this salinity.

Saline-sodic soil. Salt-affected soils with a high exchangeable sodium percentage (ESP) greater than 15%, pH usually less than 8.5.

Salt-affected soil. Soil that has been adversely affected by the presence of soluble salts, with or without high amounts of exchangeable sodium. See also *saline soil, saline-sodic soil,* and *sodic soil.*

Sand. Soil particles between 0.05 mm and 2 mm, with low specific area and also used as a texture class name for coarse soil materials. Unlike clays, sandy soils do not shrink and swell on drying and wetting and, unless artificially compacted, are rapidly permeable.

Sanitary water. Water discharged from restrooms, showers, food preparation facilities, or other non-Industrial operations; also known as 'gray water'.

Secondary mineral. A mineral resulting from the decomposition of a primary mineral or from the reprecipitation of the products of decomposition of a primary mineral.

Sediment. Particles and/or clumps of particles of sand, clay, silt, and plant or animal matter carried in water.

Sedimentary rock. A type of rock formed by chemical precipitation or by sedimentation and cementation of mineral grains transported to a site of deposition by water, wind, or ice.

Sedimentation. Deposition of sediment.

Septic tank. An underground tank to collect wastes from homes that are not connected to a municipal sewer system. Waste goes from the home to the tank and is decomposed by bacteria. Solids and dead bacteria settle to the bottom as sludge while the liquid portion flows into the ground through drains. While properly placed and maintained, septic systems can effectively treat domestic wastewater, others are a major source of ground water and surface water pollution.

Shrinking. Heavy clay soils are more susceptible to cracks formation than loamy soils, whereas in sandy soils cracks do not form or they are very small and unstable. Soil cultivation destroys crack system, mainly by tillage.

Siderochromes. Compounds produced by microorganisms that are involved with the uptake of iron by those microorganisms.

Siderophores. *See* Siderochromes.

Silt. Soil particles between 0.002 mm and 0.05 mm, with high or medium-high specific area influencing stability of soil structure; also used as a texture class name for medium and medium-fine soil materials.

Smog. Dust, smoke, or chemical fumes that pollute the air and make hazy, unhealthy conditions (literally, the word is a blend of moke and fog). Automobile, truck, bus, and other vehicle exhausts and particulates are usually trapped close to the ground, obscuring visibility and contributing to a number of respiratory problems.

Sodic soil. Soil with excess of sodium, pH is higher than 7, usually in the range 8.0 – 10.0, exchangeable sodium percentage,

ESP> 15 and very poor soil structure. These soils need special management for agriculture; non-sodic soils are without excess of sodium.

Soil biology. A scientific discipline dealing with living components of soils, which are represented mainly by bacteria, fungi, protozoa, nematodes, arthropods and earthworms as well as by mammals.

Soil chemistry. A scientific discipline dealing with chemical properties of soils and studies on the influence of fertilizers, pesticides and the other chemical substances applied on or into the soil on soil behaviour and fertility.

Soil classification. Also termed soil taxonomy, is the scientific discipline dealing with grouping of soils into soil morphological units or soil types, according to similar or comparable soil forming properties. Many countries in the world have national soil classification systems but those of FAO and USDA are used internationally. For transnational comparisons, an international soil classification system, into which the majority of national systems can be translated, is needed.

Soil colour. Soil colour is one of the indicators of soil status and depends on many factors, mainly on the amount and state of organic matter and iron oxide, as well as amount of air and water in soil pores. In general, dark soils have high organic matter content, grey soils are waterlogged or anaerobic, brown soils are well-drained and aerated soils. Soil colour is measured using *Munsell Soil Colour* charts.

Soil compaction. Changing the nature of the soil such that there is a decrease in the volume of voids between soil particles or aggregates. It is manifest as an increase in bulk density and a severely compacted soil can become effectively impermeable. Some soils are naturally compacted, e.g. very heavy textured soils (fine textured). Man-made compaction is caused by the passage of heavy machinery and very intensive soil exploitation.

Soil cracks. Openings in horizontal (mm or several cm) and vertical (cm or several m) orientation, mainly affecting soil hydraulic properties, arising from swelling and shrinking of soil.

Soil degradation. Negative process often accelerated by human activities (improper soil use and cultivation practices, building areas) that leads to deterioration of soil properties and functions or destruction of soil as a whole, e.g. compaction, erosion, salinisation.

Soil depth. Depth of soil profile from the top to parent material or bedrock or to the layer of obstacles for roots. It differs significantly for different soil types. It is one of basic criterions used in soil classification. Soils can be very shallow (less than 25 cm), shallow (25 cm-50 cm), moderately deep (50 cm-90 cm), deep (90cm-150 cm) and very deep (more than 150 cm).

Soil fertility. A measure of the ability of soil to provide plants with sufficient amount of nutrients and water, and a suitable medium for root development to assure proper plant growth and maturity.

Soil geography. Scientific discipline dealing with distribution of soil types in landscapes, describing this distribution according to geographical rules.

Soil microorganisms. Represented by protozoa, viruses, bacteria, fungi and algae. The most prevalent are bacteria and fungi, and depending on conditions (water and nutrients content, temperature, etc.) they can be in an active or non-active state. According to nutrient (and oxygen) demand, microorganisms are divided to autotropic and heterotrophic, (aerobic and anaerobic) groups. Microorganisms are a good indicator of soil status and quality.

Soil monitoring. Repeated observation and measurement of selected soil properties and functions, mainly for studying changes in soil conditions.

Soil morphology. Form and arrangement of pedological features.

Soil physics. Scientific discipline dealing with physical properties of soil (density, porosity, water retention and permeability, hydraulic conductivity etc.).

Soil porosity. Volume of water and air that can be held in a soil; ratio of the volume of voids to the total volume of the soil.

Soil profile. Vertical section of soil horizons from upper layer to the parent material, showing the arrangement (configuration) of soil horizons typical for single soil types and used as a basis for soil classification.

Soil sorption. Selective process, which occurs on soil particles smaller than 0.002mm (< 2μm). These small particles have colloidal properties, are able to hold and exchange ions, water or gases.

Soil texture. Numerical proportion (% by wt.) of sand, silt and clay in a soil. Sand, silt and clay content are estimated in the field, and/or quantitatively in the laboratory, and then placed within the texture triangle to determine soil texture class. Texture can be coarse (sand particles predominate), medium (silt particles predominate), or fine (clay particles predominate).

Solid waste management facility. Any disposal or resource recovery system; any system, program, or facility for resource conservation; any facility for the treatment of solid wastes.

Solid waste. Any solid, semi-solid, liquid, or contained gaseous materials discarded from industrial, commercial, mining, or agricultural operations, and from community activities. Solid waste includes garbage, construction debris, commercial refuse, sludge from water supply or waste treatment plants, or air pollution control faciiities, and other discarded materials.

Source control. A practice, method, or technology used to reduce pollution from a source such as best management practices.

Source reduction. The design, manufacture, purchase, or use of materials (such as products and packaging) to reduce the amount or toxicity of garbage generated. Source reduction can help reduce waste disposal and handling charges because the costs of recycling, municipal composting, landfilling, and combustion are avoided. Source reduction conserves resources and reduces pollution.

Source separation. Organizing materials by type (such as paper, metal, plastic, and glass) so that these items can be recycled instead of thrown away. For example, many of us separate these items from the rest of our household and office wastes. Industries also organize materials in this fashion.

Spores (bacterial endospores). A metabolically dormant state of bacteria in which they are more resistant to heat, chemicals, etc.

Stable isotope. Naturally occurring isotope of no observable radioactivity.

Stoniness. It is the relative proportion (volume %) of coarse particles (larger than 2 mm diameter) in the soil or on soil surface; 15% stones is a high value and can hinder cultivation and reduce water holding capacity.

Subbasins. One of several basins that form a watershed.

Subirrigation. Utilization of subsurface drainage to irrigate crops through the use of capillary rise of water from the full drainage tiles upward through the soil into the rooting zone.

Subsurface drainage. The use of tiles or plastic tubing that is buried 45 to 75 cm deep at a spacing of 8 to 30 m to drain the soil.

Subwatershed. A drainage area within a watershed.

Surface drainage. The use of open ditches that provide predominately surface drainage.

Surface soil. The layer of soil occurring on the surface, synonym *topsoil.*

Surface water. All water naturally open to the atmosphere (rivers, lakes, reservoirs, ponds, streams, seas, estuaries) and all springs, wells, or other collectors directly influenced by surface water.

Surfactant. A natural or synthetic chemical that promotes the wetting, solubilization, and emulsification of various types of organic chemicals.

Suspended load. Sediment that is transported by suspension in the water column of a stream or river.

Suspended solids. Organic and inorganic particles, such as solids from wastewater, sand, clay, and mud that are suspended and carried in water.

Sustainable agriculture. Environmentally friendly methods of farming that allow the production of crops or livestock without damage to the farm as an ecosystem, including effects on soil, water supplies, biodiversity, or other surrounding natural resources. Terms often associated with farms that are self-sustaining include *'low-input,'* organic, *'ecological,' 'biodynamic,'* and *'permaculture.'*

Swamp. Wetland that features permanent inundation of large areas of land by shallow bodies of water, generally with a substantial number of hummocks, or dry-land protrusions. Swamps are usually regarded as including a large amount of woody vegetation.

Swelling and shrinking. Two opposite processes of soil volume change. Swelling, increase of soil volume, shrinking, and decrease of soil volume. These processes are influenced by actual water content and presence of clay minerals, which are able to take or to lose water in their interlayer spaces. Difference in volume can range from 5% to more than 100% depending on quality and quantity of clay minerals.

Synergism. The cooperative action of two or more organisms producing a greater total result than the sum of their independent effects; chemicals or muscles in synergy enhance the effectiveness of one another beyond what an individual could have produced.

T

Targeting. The process of prioritizing pollutant sources for treatment with BMPs (Best Management Practices) or a specific BMP to maximize the water quality benefits of the implemented BMPs.

Teratogen. A substance capable of causing birth defects.

Thermophile. Any organism that has an optimum growth temperature above 45 degrees C.

Tidal flats. Nearly flat areas, periodically covered by tidal (periodical) waters, not suitable for agricultural use.

Till. Unsorted glacial sediment. Glacial drift is a general term for the coarsely graded and extremely heterogeneous sediments of glacial origin

Tillage. *see* ploughing.

Tolerance. Permissible residue level for pesticides in raw agricultural produce and processed foods. Whenever a pesticide is registered for use on a food or feed crop, a tolerance must be established. National Government establishes the tolerance levels, which are enforced by the Food and Drug Administration and the Department of Agriculture.

Tonnage. The amount of waste that a landfill accepts, usually expressed as tons per month. The rate at which a landfill accepts waste is limited by the landfill's permit.

Topsoil. The surface soil horizon (A) which is modified when cultivated, and designated Ap. See also surface soil. This is fertile soil material used to topdress roadbanks, gardens, and lawns.

Total Kjeldahl nitrogen. An oxidative procedure that converts organic nitrogen forms to ammonia by digestion with an acid, catalyst, and heat.

Total suspended solids (TSS). The weight of particles that are suspended in water. Suspended solids in water reduce light penetration in the water column, can clog the gills of fish and invertebrates, and are often associated with toxic contaminants because organics and metals tend to bind to particles. Total suspended solids are differentiated from total dissolved solids by a standardized filtration process, the dissolved portion passing through the filter.

Toxic chemical. Substances that can cause severe illness, poisoning, birth defects, disease, or death when ingested, inhaled, or absorbed by living organisms.

Toxic cloud. An airborne mass of gases, vapours, fumes, or aerosols of toxic materials.

Tracer. A substance, e.g. isotope, which is mixed with or fixed to the tracee (test element or compound) to follow its translocation or identify its location.

Transport. The movement of a soil particle, nutrient, or pesticide from its original position. This movement may occur in water or air currents. Nutrients and pesticides can be transported to soil particles or dissolved in water.

Tributary. A stream or river that flows into a larger stream or river.

Turbidity. A measure of the amount of light intercepted by a given volume of water due to the presence of suspended and dissolved matter and microscopic biota. Increasing the turbidity of the water decreases the amount of light that penetrates the water column. High levels of turbidity are harmful to aquatic life.

U

Unsaturated flow. Underground water flow through soil or rock where the void spaces are filled both with water and air.

V

Vadose water. Water in the vadose zone.

Vadose zone. The aerated region of soil above the permanent water table. Unsaturated zone of soil above the groundwater, extending from the bottom of the capillary fringe all the way to the soil surface.

Vegetative filter strips. A strip or area of vegetation for removing sediment, organic matter, and other pollutants from runoff and wastewater.

Viable. Living, or capable of growth.

Volatile. Any substance which evaporates quickly.

Volatilization. The transport of a liquid substance by vaporization.

W

Water management. The practice of limiting the amount of water used in activities such as animal waste flushing systems or milking operations in order to reduce the amount of runoff and, therefore, decrease the probability of polluting nearby surface water.

Water quality standards. Established limits of certain chemical, physical, and biological parameters in a water body. Water quality standards are established for the different designated uses of a water body.

Water retention. The ability of soil to hold water for a period that is longer than infiltration, normally 48hr in a freely draining soil. It strongly depends on organic matter and bulk density. Soil texture also has an influence on water retention.

Water table management systems. The practices of surface drainage, controlled drainage, or sub-irrigation that are used in combination to control water table depth. Used in conjunction, these practices can increase yield and decrease nutrient pollution.

Water table. The upper surface of ground water or that level in the ground where the water is at atmospheric pressure.

Watershed. The area of land from which rainfall (and/or snow melt) drains into a single point. Watersheds are also sometimes referred to as drainage basins or drainage areas. Ridges of higher ground generally form the boundaries between watersheds. At these boundaries, rain falling on one side flows toward the low point of one watershed, while rain falling on the other side of the boundary flows toward the low point of a different watershed.

Weathering. The breakdown and changes in rocks and sediments at or near the Earth's surface produced by biological, chemical, and physical agents or combinations of them.

Wellhead protection area. A protected surface and subsurface zone surrounding a well or well field that supplies a public water system and through which contaminants could likely reach well water.

Wetland creation. The bringing into existence of a wetland, whether by accident or intentionally, where none existed previously, for purposes including mitigation, habitat provision, and water quality improvement.

Wetland enhancement. Modification of a natural or created wetland to enhance one or more functions, typically to the detriment of other functions.

Wetland restoration. Rehabilitation of previously existing wetland functions, from a more impaired to a less impaired or unimpaired state of overall function.

Wetlands. Areas that are soaked or flooded by surface or ground water frequently enough or for sufficient duration to support plants, birds, animals, and aquatic life. Wetlands generally include swamps, marshes, bogs, estuaries, and other inland and coastal areas, and are federally protected. Wetlands frequently serve as recharge/discharge areas and are known as "nature's kidneys" since they help purify water. Wetlands also have been referred to as natural sponges that absorb flood waters, functioning like natural tubs to collect overflow. Wetlands are important wildlife habitats, breeding grounds, and nurseries because of their biodiversity. Many endangered species as well as countless estuarine and marine fish and shellfish, mammals, waterfowl, and other migratory birds use wetland habitat for growth, reproduction, food, and shelter. Wetlands are among the most fertile, natural ecosystems in the world since they produce great volumes of food (plant material).

Wildlife refuge. An area designated for the protection of wild animals, within which hunting and fishing are either prohibited or strictly controlled.

Wilting point. Soil moisture content when the rate of absorption of water by plant roots is too slow to maintain plant turgidity and permanent wilting occurs. The average moisture tension at the outside surface of the moisture film around soil particles when permanent wilting occurs is 15 atmospheres or 1500kPa.

Wood treatment facility. An industrial facility which treats lumber and other wood products for outdoor use. The process involves use of chromated copper arsenate and other toxic chemicals which are regulated as hazardous materials.

X

Xenobiotic. A term for non-natural or man-made substances found in the environment (i.e., synthetics, plastics).

Xerophile. Organism adapted to grow at low water potential, i.e., very dry habitats.

Index

T

V

W

Z